Agriculture and Food Security

Published By

Oxford Book Company

Regd. Off. "Bony Residency", Gate No. 2
Opp. Tilak Public School, Vishweswariya Nagar,
Gopalpura Road, Jaipur - 302018, India
Ph: 0141 - 2761280, Telefax: 2761381
email - oxfordbookjpr@gmail.com
website - oxfordbookcompany.com

Delhi Office: 102, 1st Floor, Satyam House,
4327/3, Ansari Road, Darya Ganj,
New Delhi - 110002, India
Ph: 011-45652440

Distributed By

KEYA INTERNATIONAL PRESS LLP
E-mail: keyainternationalpressllp@gmail.com

Published in India, 2024

ISBN : 9788189473280

Price: ₹ 1850.00

Typeset by Shivangi Computers, Jaipur

Printed and bound in India by Thomson Press Ltd.,
New Delhi

Preface

Food security exists when all people, at all times, have access to sufficient, safe and nutritious food to meet their dietary needs and food preferences for an active and healthy life. For a household, food security means access by members at all times to enough food for an active and healthy life. It includes a minimum ready availability of nutritionally adequate and safe foods. A direct relationship exists between food consumption levels and poverty. Worldwide, millions of people are chronically hungry due to extreme poverty.

Humans must have food to survive. In India and other countries, the system of obtaining food was highly localised. However, with the development of new technologies and improved transportation, food production and distribution systems became national in scope and more complex and international in approach. Ensuring its quantity, nutritional adequacy and safety have become more complicated and require major government and private efforts. Food security encompasses a wide spectrum of issues—not only the avoidance of foodborne pathogens, chemical toxicants and physical hazards, but also issues such as nutrition, food quality, labelling and education.

Worldwide, agriculture plays main role in meeting food needs of the people. However, fast increasing population, poverty among masses and insecured food are main obstacles in fulfilling the food needs of people. Present book describes basic concepts relating to food security and safety, food hazards, misbranded food, nutrition labelling, rising food insecurity, agricultural issues in food security and regulations for food safety and security. In addition, the book also describes effective food safety systems and future prospect of food.

Archana Ruhela

Contents

	Preface	*v*
1.	Introduction	1
2.	Causes and Background of Food Security	11
3.	The Regulation of Misbranded Food	20
4.	The Nature of Food Hazards	50
5.	Effective Food Safety System	64
6.	Values for Nutrition Labeling	87
7.	Food and Drug Administration	134
8.	Agriculture and Rising Food Insecurity	152
9.	Food Security and Rural Development	172
10.	The Future of Food	187
11.	War and the World's Food Economy, Agriculture, and the Rural Community	208
12.	The Environment of Agriculture	244
	Index	275

1

Introduction

FOOD SECURITY

Food security describes a situation in which people do not live in hunger or fear of starvation. World-wide around 852 million men, women and children are chronically hungry due to extreme poverty; while up to 2 billion people lack food security intermittently due to varying degrees of poverty.

A direct relationship exists between food consumption levels and poverty. Families with the financial resources to escape extreme poverty rarely suffer from chronic hunger; while poor families not only suffer the most from chronic hunger, but are also the segment of the population most at risk during food shortages and famines.

Two commonly used definitions of food security come from the UN's Food and Agriculture Organization (FAO) and the United States Department of Agriculture (USDA):

- Food security exists when all people, at all times, have access to sufficient, safe and nutritious food to meet their dietary needs and food preferences for an active and healthy life. (FAO)
- Food security for a household means access by all members at all times to enough food for an active, healthy life. Food security includes at a minimum
- The ready availability of nutritionally adequate and safe foods,

- An assured ability to acquire acceptable foods in socially acceptable ways (that is, without resorting to emergency food supplies, scavenging, stealing, or other coping strategies). (USDA)

The stages of food insecurity range from food secure situations to full-scale famine. "Famine and hunger are both rooted in food insecurity. Food insecurity can be categorized as either chronic or transitory. Chronic food insecurity translates into a high degree of vulnerability to famine and hunger; ensuring food security presupposes elimination of that vulnerability. Chronic hunger is not famine. It is similar to undernourishment and is related to poverty. It exists mainly in poor countries."

FOOD INSECURITY

Food insecurity has been described as "a condition in which people lack basic food intake to provide them with the energy and nutrients for fully productive lives." Hunger Task Force

Food insecurity is defined as not having access to enough food every day of the year. The U.S. federal government collects statistics on this based on surveys. Thus a "food insecure household" is one in which any member experienced one or more days in a given year when they didn't have enough food because of poverty or a lack of resources. This poverty measure takes into account participation in food assistance programmes such as school breakfast, school lunch and Food Stamps.

The USDA report cited below asks the question, "How often were people hungry in households that were food insecure with hunger?" Around four percent of people reported going hungry at least once a year, while on any given day the figure is estimated to be between 0.5 percent and 0.8 percent.

Chronic Nutritional Deficiencies

Many countries experience perpetual food shortages and distribution problems. These result in chronic and often widespread hunger amongst significant numbers of people. Human populations respond to chronic hunger and malnutrition by decreasing body size, known in medical terms as stunting or stunted growth. This process starts *in utero* if the mother is malnourished and continues through approximately the third year of life. It leads to higher infant and child mortality, but at rates far lower than during famines. Once stunting has occurred, improved nutritional intake later in life cannot reverse the damage. Stunting itself is viewed as a coping mechanism, designed to bring body size into alignment with the calories available during adulthood in the location where the child is born. Limiting body size as a way of adapting to low levels of energy (calories) adversely affects health in three ways:

- Premature failure of vital organs occurs during adulthood. For example a 50 year old individual might die of heart failure because his/her heart suffered structural defects during early development.
- Stunted individuals suffer a far higher rate of disease and illness than those who have not undergone stunting.
- Severe malnutrition in early childhood often leads to defects in cognitive development.

"The analysis points to the misleading nature of the concept of subsistence as Malthus originally used it and as it is still widely used today. Subsistence in not located at the edge of a nutritional cliff, beyond which lies demographic disaster. Rather than one level of subsistence, there are numerous levels at which a population and a food supply can be in equilibrium in the sense that they can be indefinitely sustained. However, some levels will have smaller people and higher normal mortality than others."

Dictatorship and Kleptocracy

As the Nobel Prize-winning economist Amartya Sen has observed that "there is no such thing as an apolitical food problem." While drought and other naturally occurring events may trigger famine conditions, it is government action or inaction that determines its severity, and often even whether or not a famine will occur. The 20th century is full of examples of governments undermining the food security of their own nations – sometimes intentionally.

Socialist anti-Western governments do this much more than authoritarian, and the Soviet Union in the mid-20th century. Lesser known are Cambodia under Pol Pot, and the Sandinistas in Nicaragua in their attempt to starve the Misquito Indians into submission.

When governments come to power by force or rigged elections, and not by way of fair and open elections, their base of support is often narrow and built upon cronyism and patronage. Under such conditions "The distribution of food within a country is a political issue. Governments in most countries give priority to urban areas, since that is where the most influential and powerful families and enterprises are usually located. The government often neglects subsistence farmers and rural areas in general. The more remote and underdeveloped the area the less likely the government will be to effectively meet its needs. Many agrarian policies, especially the pricing of agricultural commodities, discriminate against rural areas. Governments often keep prices of basic grains at such artificially low levels that subsistence producers can not accumulate enough capital to make investments to improve their production. Thus, they are effectively prevented from getting out of their precarious situation."

Further Communist parties, as well as dictators and warlords have used food as a political weapon, rewarding their supporters while denying food supplies to areas that oppose their rule. Under such conditions food becomes a currency with which to buy support and famine becomes an effective weapon to be used against the opposition.

Governments with strong tendencies towards kleptocracy can undermine food security even when harvests are good. When government monopolizes trade, farmers may find that they are free to grow cash crops for export, but under penalty of law only able to sell their crops to government buyers at prices far below the world market price. The government then is free to sell their crop on the world market at full price, pocketing the difference. This creates an artificial "poverty trap" from which even the most hard working and motivated farmers may not escape.

When the rule of law is absent, or private property is non-existent, farmers have little incentive to improve their productivity. If a farm becomes noticeably more productive than neighbouring farms, it may become the target of individuals well connected to the government. Rather than risk being noticed and possibly losing their land, farmers may be content with the perceived safety of mediocrity.

Economic Approaches

There are many economic approaches advocated to improve food security in developing countries. Three typical approaches are listed below. The first is typical of what is advocated by most governments and international agencies. The other two are more common to non-governmental organizations (NGO's).

Westernized View

Conventional thinking in westernized countries is that maximizing the farmers profit is the surest way of maximizing agricultural production; the higher a farmer's profit, the greater the effort that will be forthcoming, and the greater the risk the farmer is willing to take.

This view holds that it is the governments job to place into the hands of farmers the largest number and highest quality tools possible (tools is used here to refer to improved production techniques, improved seeds, secure land tenure, accurate weather forecasts, etc.) However, it is left to the individual farmer to pick and choose which tools to use, and how to use them, as farmers have intimate knowledge of their own land and local conditions.

As with other businesses, a percentage of the profits are normally reinvested into the business in the hopes of increasing production, and hence increase future profits. Normally higher profits translate into higher spending on technologies designed to boost production, such as drip irrigation systems, agriculture education, and greenhouses. An increased profit also increases the farmer's incentive to engage in double-cropping, soil improvement programmes, and expanding usable acreage.

Food Justice

An alternative view takes a collective approach to achieve food security. It notes that globally enough food is produced to feed the entire world population at a level adequate to ensure that everyone can be free of hunger and fear of starvation. That no one should live without enough food because of economic constraints or social inequalities is the basic goal.

This approach is often referred to as food justice and views food security as a basic human right. It advocates fairer distribution of food, particularly grain crops, as a means of ending chronic hunger and malnutrition. The core of the Food Justice movement is the belief that what is lacking is not food, but the political will to fairly distribute food regardless of the recipient's ability to pay.

Food Sovereignty

A third approach is known as food sovereignty; though it overlaps with food justice on several points, the two are not identical. It views the business practices of multinational corporations as a form of neocolonialism. It contends that multinational corporations have the financial resources available to buy up the agricultural resources of impoverished nations, particularly in the tropics. They also have the political clout to convert these resources to the exclusive production of cash crops for sale to industrialized nations outside of the tropics, and in the process to squeeze the poor off of the more productive lands. Under this view subsistence farmers are left to cultivate only lands that are so marginal in terms of productivity as to be of no interest to the multinational corporations.

It advocates banning the production of most cash crops in developing nations, thereby leaving the local farmers to concentrate on subsistence crops. In addition it opposes allowing low-cost subsidized food from industrialized nations into developing countries, what is referred to as "import dumping".

World Food Summit

The World Food Summit was held in Rome, with the aim of renewing global commitment to the fight against hunger. The Food and Agriculture Organization of the United Nations (FAO) called the summit in response to widespread under-nutrition and growing concern about the capacity of agriculture to meet future food needs. The conference produced two key documents, the Rome Declaration on World Food Security and the World Food Summit Plan of Action.The Plan of Action sets a number of targets for government and non-governmental organizations for achieving food security, at the individual, household, national, regional and global levels.

Achieving Food Security

"The number of people without enough food to eat on a regular basis remains stubbornly high, at over 800 million, and is not falling significantly. Over 60 per cent of the world's undernourished people live in Asia, and a quarter in Africa. The proportion of people who are hungry, however, is greater in Africa (33 per cent) than Asia (16 per cent). The latest FAO figures indicate that there are 22 countries, 16 of which are in Africa, in which the undernourishment prevalence rate is over 35 per cent." – Food and Agriculture Organization (FAO).'In general the countries that succeeded in reducing hunger were characterised by more rapid economic growth and specifically more rapid growth in their agricultural sectors. They also exhibited slower population growth, lower levels of HIV and higher ranking in the Human Development Index'.

USAID proposes several key steps to increasing agricultural productivity which is in turn key to increasing rural income and reducing food insecurity. They include:

- Boosting agricultural science and technology. Current agricultural yields are insufficient to feed the growing populations. Eventually, the rising agricultural productivity drives economic growth.
- Securing property rights and access to finance.
- Enhancing human capital through education and improved health.
- Conflict prevention and resolution mechanisms and democracy and governance based on principles of accountability and transparency in public institutions and the rule of law are basic to reducing vulnerable members of society.

The First Millennium Development Goal is to Eradicate extreme hunger and poverty, and agricultural productivity is likely to play a key role if it is to be reached on time.

"Of the eight Millennium Development Goals, eradicating extreme hunger and poverty depends on agriculture the most.

The Agriculture-hunger-poverty Nexus

Eradicating hunger and poverty requires an understanding of the ways in which these two injustices interconnect. Hunger, and the malnourishment that accompanies it, prevents poor people from escaping poverty because it diminishes their ability to learn, work, and care for themselves and their family members. Food insecurity exists when people are undernourished as a result of the physical unavailability of food, their lack of social or economic access to adequate food, and/or inadequate food utilization. Food-insecure people are those individuals whose food intake falls below their minimum calorie (energy) requirements, as well as those who exhibit physical symptoms caused by energy and nutrient deficiencies resulting from an inadequate or unbalanced diet or from the body's inability to use food effectively because of infection or disease.

An alternative view would define the concept of food insecurity as referring only to the consequence of inadequate consumption of nutritious food, considering the physiological utilization of food by the body as being within the domain of nutrition and health. Malnourishment also leads to poor health hence individuals fail to provide for their families. If left unaddressed, hunger sets in motion an array of outcomes that perpetuate malnutrition, reduce the ability of adults to work and to give birth to healthy children, and erode children's ability to learn and lead productive, healthy, and happy lives. This truncation of human development undermines a country's potential for economic development – for generations to come.

There are strong, direct relationships between agricultural productivity, hunger, and poverty. Three-quarters of the world's poor live in rural areas and make their living from agriculture. Hunger and child malnutrition are greater in these areas than in urban areas. Moreover, the higher the proportion of the rural population that obtains its income solely from subsistence farming (without the benefit of pro-poor technologies and access to markets), the higher the incidence of malnutrition. Therefore, improvements in agricultural productivity aimed at small-scale farmers will benefit the rural poor first.

Increased agricultural productivity enables farmers to grow more food, which translates into better diets and, under market conditions that offer a level playing field, into higher farm incomes. With more money, farmers are more likely to diversify production and grow higher-value crops, benefiting not only themselves but the economy as a whole."

Biotechnology for Smallholders

The area sown to genetically enhanced crops in developing countries is rapidly catching-up with the area sown in industrial nations. According to the International Service for the Acquisition of Agri-biotech Applications (ISAAA), genetically enhanced (biotech, GM) crops were grown by approximately 8.5 million farmers in 21 countries in 2005, up from 8.25 million farmers in 17 countries in 2004. The largest increase in biotech crop area in any country in 2005 was in Brazil, provisionally estimated at 44,000 km^2 (94,000 km^2 in

2005 compared with 50,000 km^2 in 2004. India had by far the largest year-on-year proportional increase, with almost a threefold increase from 5,000 km^2 in 2004 to 13,000 km^2 in 2005.

Current high regulatory costs imposed on varieties created by the more modern methods are a significant hurdle for development of genetically enhanced crops well suited to developing country farmers by modern genetic methods. Once a new variety is developed, however, seed provides a good vehicle for distribution of improvements in a package that is familiar to the farmer.

Currently there are some institutes and research groups that have projects in which biotechnology is shared with contact people in less-developed countries on a non-profit basis. These institutes make use of biotechnological methods that do not involve high research and registration costs, such as conservation and multiplication of germplasm and phytosanitation.

2

Causes and Background of Food Security

BACKGROUND

Humans must have food to survive. In India and other countries, the system of obtaining food was highly localized before the twentieth century. With the development of new technologies and improved transportation, food production and distribution systems became national in scope and more complex. The current food system stretches from producers to consumers and is international in scope. Ensuring its quantity, nutritional adequacy, and safety has become more complicated, and requires major government and private-sector efforts.

Food safety encompasses a wide spectrum of issues-not only the avoidance of foodborne pathogens, chemical toxicants, and physical hazards, but also issues such as nutrition, food quality, labeling, and education. The system for regulating the food supply in the United States involves all levels of government from federal to local. The present legal framework is comprised of many inconsistent statutes and regulations, and implementing authority is spread among at least 12 federal agencies. Such a fragmented structure requires heroic efforts at cooperation, communication, and coordination (federal agencies have reported more than 50 interagency agreements), but duplication of efforts and regulatory gaps are common. Food safety problems that transcend the jurisdictional boundaries of two or more agencies are often not reported in the most expeditious manner. The General Accounting Office (GAO) reports that the National Marine Fisheries Service (NMFS), which operates a voluntary seafood inspection programme, failed to notify the Food and Drug Administration (FDA), the agency with regulatory responsibility for corrective action, of 198 plants that failed NMFS inspection.

Recent outbreaks involving such items as Guatemalan raspberries, hamburger, ice cream, and cereal have raised concern over the adequacy of the current system to ensure the safety of the US food supply. The GAO, public interest groups, and several members of Congress have suggested the consolidation of the existing federal food safety structure into a single food safety agency. In addition, the Institute of Food Technologists (IFT) recently developed *Guiding Principles for Optimum Food Safety Oversight and Regulation in the United States,* which describes attributes of an effective food safety system. These principles have been endorsed by 13 professional, scientific societies.

To improve the safety of the US food supply, President Clinton directed the Secretary of Agriculture, the Secretary of Health and Human Services, and the Administrator of the Environmental Protection Agency to develop a food safety initiative. *Food Safety from Farm to Table: A National Food-Safety Initiative* seeks to address hazards that present the greatest risk, make the best use of public and private resources, increase collaboration between public and private organizations, and improve coordination in the government. Recently, President Clinton announced a plan to create a Joint Institute for Food Safety Research that will develop a coordinated strategy for conducting food safety research consistent with the above national initiative.

CHANGES IN THE FOOD SYSTEM

The Indian food supply is abundant and affordable and it is acknowledged by many to pose an acceptable level of risk. The food system has evolved from one that provided consumers with minimally processed basic commodities for home meal preparation to today's availability of highly processed products that are ready-to-eat or require minimal preparation. Food preservation processes have changed dramatically from traditional salting, curing, drying, and heating. Today's products are the result of many technological developments such as pasteurization, irradiation, and genetic engineering. Likewise, food distribution systems have changed greatly. The broad introduction of refrigerated railcars and trucks, freezers, and air transport created a national and now global food system. These changes allow out-of-season availability, convenience, variety, and improved sensory attributes.

The size and complexity of the Indian food system raise many safety issues, for example:

- The emergence of new foodborne pathogens such as *Escherichia coli* 0157:H7 and the re-emergence of previously identified pathogens such *as Salmonella* have resulted in new microbiological hazards.
- Advances in science and technology that allow the development of genetically modified foods and the construction of modified macronutrients require new ways of evaluating the safety of substances added to the food supply, and this need will increase.
- Heightened consumer interest in raw or minimally processed fruits and vegetables, partly in response to dietary recommendations, has created a year-round demand for fresh produce, which can be met out-of-season only through increased imports. This increased produce volume requires additional attention to possible contamination of domestic as well as imported fruits and vegetables.
- Americans eat fewer home-prepared meals than ever before, in response to changes in the US workforce and to developments in food processing and food service that offer greater convenience and variety in available foods. The potential for contamination increases as food is handled by more people.

- As technology has advanced, a smaller number of food facilities provide food to larger numbers of Indian consumers, increasing the extent of harm that can arise from a single incident.
- The remarkable success of modern medicine in extending the lifespan and increasing the quality of life for many people has placed new demands on the food system and on those responsible for guarding its safety. Increasing numbers of people have immune systems that are compromised because of age, illness, or medical treatment. These people are highly susceptible to illness and death from microbial pathogens, and might be more sensitive to new food ingredients and recently identified natural components of the diet.

  Increasing consumption of fortified foods and dietary supplements, including herbals, has raised new questions about the safety of "natural" substances not normally in the diet, or normally part of the diet but at much lower concentrations, and about the health effects of consuming high concentrations of nutrients ordinarily considered safe.

Thus, the developments that have provided the American consumer with a wide array of food products have also introduced risks. Government has attempted to address such developments by adding structures and processes without always considering their effects on other aspects of the system. As a result, inefficiencies and apparent conflicts within the system have arisen. Some have been corrected (for example, pesticides have been exempted from the Delaney clause's ban on carcinogens), but others remain. For example, inspectors from multiple agencies oversee parallel and nearly identical processes in the same food processing facility.

The current food safety system has evolved piecemeal over almost a century in response to changes in the food supply and to changes in the biophysical and social environments in which the system operates. The present system is not the product of planning, and it is often not equipped to anticipate changes. But the situation is not just haphazard; changes in risks have made the system outmoded. The role and organization of government entities have remained largely unchanged, and the food safety system has fallen behind today's needs.

Scope of The Food Safety Problem

Over the years, as the agriculture and marketing systems have evolved to provide food to a growing and increasingly sophisticated population, complex processes built on advances in science and technology have been developed to evaluate and manage the risks associated with the food supply. In spite of well-established systems that control many food risks, serious hazards to public health remain. Although estimates vary greatly, there is agreement that foodborne illness is a serious problem. In the United States, an estimated 81 million cases of foodborne illness may occur each year, resulting in as many as 9,000 deaths. The estimated annual medical costs and productivity losses due to seven major foodborne pathogens range from $6.6 billion to $37.1 billion.

The responsibility for managing foodborne risks is shared throughout the system because the wholesomeness and safety of a food are influenced by all the people and processes that handle or transform it from production to consumption. Federal, state, and local governments play major roles in managing risks to protect the public from hazards in the food supply. Regulatory agencies are empowered to prescribe rules, standards, and processes to control risks; to develop and maintain research programmes to apply contemporary science and technology to safety decisions; to monitor risks in the food supply; and to provide information and education to all components of the food system. In authorizing and funding efforts to ensure the safety of the food supply, the government must balance the interests of diverse groups and allocate finite resources among competing needs.

Faulty handling of a product at any point in the system can transform a safe product into one that can cause serious harm. Producers, shippers, importers, processors, wholesalers, retailers, handlers, and consumers all influence the health risks associated with food products. Consumers play a particularly important role in the control of microbiological risks, both in their food handling practices and in their demand for an effective, efficient food safety system.

Food safety issues also involve the interplay of domestic and international legal, political, scientific, social, and economic forces. Intense debates about desired levels of protection and about the appropriateness of different control measures can arise as parties discuss the scientific bases for risk decisions, public expectations, and relative costs and benefits of intervention in different ways and in different components of the food system.

Science must play a vital role in food safety decisions through risk assessment, that is, the identification of hazards and the determination of the likelihood and severity of risk under given conditions of exposure. Hazards are biological, chemical, or physical substances that can cause adverse consequences. Hazards associated with food include microbiological pathogens, naturally occurring toxins, allergens, intentional and unintentional additives, modified food components, agricultural chemicals, environmental contaminants, animal drug residues, and excessive consumption of some dietary supplements. In addition, certain methods of food preparation in the home can contribute to increasing some of these hazards. However, a hazard does not pose a risk in the absence of exposure. The susceptibility of the consumer and the magnitude of exposure determine whether those hazards cause immunological changes, genetic and developmental changes, cancer, or death.

Hazard identification is the basis for estimating risk. A food safety risk is the probability of harm to health resulting from a food-related hazard at a particular exposure to a specified person or group. It is important to recognize that safety is an intellectual concept, not an inherent biological property of a substance; safety has been defined as the judgment of an acceptable level of risk. Thus, "safe food" involves a subjective evaluation of social issues and values, as well as a scientific assessment of risk.

Food Safety Regulation

Regulation of food safety was largely the responsibility of state and local officials until the first decade of the twentieth century. The emergence of a national market for food, combined with shocking stories of practices in the food industry, spurred the federal government to reassess its responsibility to ensure food safety.

Although those seminal laws provide the basic legal and institutional framework for the federal regulation of food safety that we observe today, they are very different in their implementation and responsibilities. Their core was similar: each prohibited the shipment in interstate commerce of food that met any of several definitions of "adulteration." Those definitions, or proscriptions, targeted products that were spoiled, contaminated with filth, derived from diseased animals, or contained unsafe substances. In short, both laws sought to prevent the sale of food that could offend consumers or cause them to become ill.

The original Meat Inspection Act and Pure Food and Drugs Act focused on different sectors of the food market and, from the beginning, supported different approaches to ensuring compliance. The Meat Inspection Act, which has been administered by the US Department of Agriculture (USDA) since its inception, requires continuous on-site factory inspections by government inspectors using sight, smell, and touch to detect problems. The USDA also administers the Poultry Products Inspection Act, which requires continuous inspection of poultry and poultry products and the 1970 Egg Product Inspection Act, which requires continuous inspection of the processing of liquid, frozen, and dried egg products. Animal carcasses and processed products cannot be shipped from the processing plant until they have been inspected and deemed to be unadulterated by the USDA inspector.The meat and poultry inspection system still requires sight, smell, and touch inspections, which are ineffective in addressing the current issues of food safety. Expert opinion that USDA inspection procedures do not adequately address the microbiological problems associated with meat and poultry led to the adoption of a new approach to inspection using the Hazard Analysis Critical Control Point (HACCP) system. In contrast to the reactive characteristics of traditional food safety strategies, the HACCP system focuses on preventing hazards that could cause foodborne illness by applying science-based controls at each step of the process from raw material to finished product.

Agencies other than USDA and FDA-such as the National Marine Fisheries Service (NMFS) in the Department of Commerce, the Environmental Protection Agency (EPA), the Centres for Disease Control and Prevention (CDC), and the National Institutes of Health (NIH)-also have roles in food safety. NMFS conducts a voluntary seafood inspection and grading programme and does research on seafood safety. The EPA Office of Prevention, Pesticides, and Toxic Substances is responsible for risk assessment, product approvals, tolerance setting, and research on pesticide residues in or on human food and animal feed. State health departments in conjunction with CDC are responsible for the surveillance and investigation of illnesses related to foods, and NIH conducts research related to foodborne disease. Many state and local agencies are also involved in food safety regulation and inspection.

With so many agencies involved in food safety, decisions and priorities often focus on specific issues rather than strategies, and they are not always well-integrated. The system that has evolved in the federal government for regulating food safety is complex, fragmented, and cumbersome.

The Committee And Its Charge

Congress directed USDA's Agricultural Research Service to contract with the National Academy of Sciences to conduct a study on the scientific and organizational needs of an effective food safety system. The Academy appointed an ad hoc Food Safety Oversight Commission composed of selected members of the Institute of Medicine's Food and Nutrition Board, the National Research Council's (NRC) Board on Agriculture, and the NRC Commission on Life Science's Board on Environmental Studies and Toxicology; in turn, the ad hoc commission convened the Committee to Ensure Safe Food from Production to Consumption.

The committee was charged with determining the scientific basis of an effective food safety system; assessing the effectiveness of the current US system to ensure safe food; identifying scientific and organizational needs and gaps in the current system at the federal level; and providing recommendations to move toward the scientific foundation and organizational structure of a more effective food safety system for present and future generations. If organizational changes were recommended, a subsequent study on implementing the recommendations was to follow, if requested. Given the constraints on time and funding, the committee focused on the role of the federal government in ensuring safe food.

The committee is aware that safe food is and always will be a moving target with respect to both changing hazards and changing degrees of acceptance of risk. This report is not and cannot be a comprehensive, all-inclusive discussion of these issues and adoption of its recommendations will not end the effort to make food safer. It is not possible to foresee how risk acceptance will change or how the problems will change in decades to come, but federal food safety efforts must be designed effectively to deal with what is known and what is not known.

3

The Regulation of Misbranded Food

CONTEXTUAL FACTORS

Any consideration of reforming the intergovernmental regulatory framework governing food labeling requires a basic understanding of the historical development of food regulation in general, the evolution of food manufacturing and marketing, and the impact of major public health events in India since the turn of the century. In addition, understanding the background to the current regulatory environment since passage of the Federal Food, Drug, and Cosmetic Act (FDCA) helps to identify the rationale for existing statutes and regulations in various jurisdictions and the need for reform.

Changes in food manufacturing and marketing have led to the increasing nationalization of the food supply, which has been accompanied by local economic protection of specific food commodities. Efforts to protect the public's health, have also increased the complexity of food laws and regulations. These changes also may be linked to determining the appropriate level of government at which to regulate food labeling. In some cases, a level of expertise or efficiency may be required that only the Food and Drug Administration (FDA) possesses. Merchandising practices, packaging, or other features of food marketing may require national rather than regional control. The nation's system of food regulation has, in effect, resulted as a response from State and Federal regulators to changes in food manufacturing and marketing and concerns about public health and consumer protection over time.

DEVELOPMENTS BEFORE 1900

Food Production and Marketing

Early in the nineteenth century, the development of shelf-stable food products in hermetically sealed containers was a major landmark in the history of food packaging. The use of tin cans to actually seal and cook a food began in England in the early 1800s. This convenient food package was a major step forward in packaging, providing a year-round food supply in three-piece sanitary containers with processing built in and preparation requiring only reheating and/or adding water.

In the late 1800s, several innovations in packaging had developed as a result of the industrial revolution, including the metal can for heat-processed foods described above, the collapsible tube, the folding carton, the corrupted shipping case, and the crown closure (for hermetically sealing narrow-neck bottles). These developments led to the introduction and commercialization of the milk battle and canned condensed milk, which together with pasteurization had a positive impact on public health and infant mortality. The Uneeda Biscuit package provided a consumer-sized quantity of crackers, undoubtedly leading to the self-serve era.

At that time, bulk dry goods, such as flour and sugar, were sold in country and general stores. As the nation approached the twentieth century, specialty stores developed, with consumers making separate stops at the butcher shop, the bakery, and the produce stand—that is, if horse-and-wagon peddlers did not provide the items needed at the consumer's doorstep. The small retail food specialty store was run independently by the owner, and most items came in bulk form.

The Great Atlantic and Pacific Tea Company, better known as A&P, is generally credited with creating the first chain grocery stores—the forerunner of the modern supermarket. Established during the mid-1800s, early A&P stores resembled a typical specialty store: one person ran the whole operation, standing behind the counter and handing customers their orders from the shelves. However, in contrast to specialty stores, these chain operations had a direct link with suppliers, which allowed them to charge consumers lower prices than their competitors.

Public Health

The importance of public health measures applied to the community at large in reducing rates of illness and death cannot be overemphasized. During the nineteenth century, the isolation of illness was a significant step forward in limiting the spread of disease. In addition, expanding knowledge of bacteriology led to improvements in community cleanliness in the form of water purification and sewage disposal systems. The pasteurization of milk represented significant progress for the safety of this commodity, with related sanitary measures being instituted in the manufacturing, packing, and sale of other foods.

In the 1800s, a series of publications documented adulteration of the food supply and the resulting negative impact on both the nation's economy and public health. In 1850, a landmark public health report by Shattuck documented the decrease in average life expectancy at birth in America's large urban centres and identified the adulteration of food and drugs as a matter of public health concern. The report recommended that local boards of health be established to "endeavor to prevent the sale and use of unwholesome, spurious, and adulterated articles, dangerous to the public health, designed for food, drink, or medicine". Following that report, boards of health were established by cities, counties, and States throughout the country. State Departments of Agriculture were also established with authority to regulate the manufacture of food products. In 1888, Congress enacted the first broad food and drug legislation for the District of Columbia, which was subsequently strengthened in 1893.

Food Regulation

Following the American Revolution, the nation was governed first under the Articles of Confederation, which permitted the States to regulate activities within their borders. However, the Articles ultimately failed to ensure a cohesive union because each State retained its own sovereign powers, leaving the national government without the authority to govern or resolve either domestic or foreign problems.

A compromise made during the Constitutional Convention of 1789 allowed the States to retain their traditional powers, including the "police powers" under which they could act to protect the health and welfare of the public. At the same time, the Convention agreed that laws enacted by the Federal government under its enumerated powers would be supreme, including the laws regulating "commerce among the several States". The doctrine of Federal preemption originated with the supremacy clause of the Constitution, under which the laws of the Federal government "shall be the supreme law of the land," giving them priority over any State or local law. Under this clause, conflicts between Federal laws and State requirements (i.e., statutes and regulations) are settled by making the Federal law preemptive of those of the States. This power applies both to State provisions that are in direct conflict with Federal laws and those State laws that interfere with Congressional objectives. Under the commerce clause, the courts have also struck down inconsistent State laws for placing an unreasonable burden on interstate commerce.

To those who opposed a strong central government and feared the demise of State governments at the time of the Constitutional Convention, James Madison argued that, for a number of practical reasons, including proximity to the people, State governments would not only survive but would remain a vital component of the Federal system. Madison further argued that the real issue was not the inherent correctness of granting power to a particular level of government but how best to carry out the will of the people. The purpose of government at both the Federal and State levels was to do the peoples bidding, and when the majority of the populace spoke through the Federal legislature, the prerogatives of the State governments must be superseded.

Actual implementation of Federal preemption has been more complicated than the founding fathers envisioned, in part because of the early recognition that certain functions of government are local in nature. The preemption doctrine constitutes a recognition that Congress allocates decision making power to different levels of government, either by assuming the power for the Federal government or leaving it to the States. Frequently, Congress is unclear in this delineation, which calls for a decision to be made by the courts. Under modern interpretation of the commerce clause, Congress had plenary authority to regulate food labels. If it chose, it could oust all State regulation of any label on any food. Ordinarily, it has not chosen this option, leaving States considerable freedom to regulate foods sold within their borders, including foods shipped out of State, so long as their requirements do not unreasonably burden interstate commerce.

Consequently, during the 1800s, the Federal government addressed food problems surrounding imports and exports with statutes enacted by Congress to regulate foreign commerce in food as commercial food production and markets grew. States continued to regulate local activities of the food industry, such as retail food sanitation. By the turn of the century, Harvey Wiley was conducting studies on adulteration and labeling in the Division of Chemistry of the U.S. Department of Agriculture (USDA). However, there was no Federal statutory authority to regulate the interstate sale of adulterated or misbranded foods.

It is very certain that the widely differing statutes relating to our food supply in the different States have worked much mischief, been the cause of much confusion, and seriously embarrassed some useful industries. I think all who have studied the matter will be inclined to admit that uniformity in our food laws is much to be desired.

Several organizations had been formed with at least one goal being to work toward the establishment of uniformity. The original 1884 constitution of the Association of Official Agricultural Chemists (which subsequently became the Association of Official Analytical Chemists) stated that the objectives of the organization were "to secure, as far as possible, uniformity in legislation and uniformity and accuracy in the methods and results" of analysis. In 1897, representatives from 10 States met "for the purpose of forming a national Association with the end in view of producing, as nearly as conditions and laws would permit, uniformity of action in the enforcement of such food and drug laws". The Constitution adopted by the resulting organization, the Association of Food and Drug Officials (AFDO), stated that the group's purpose was "to promote and foster such legislation as would tend to protect public health and prevent deception also to promote uniformity in legislation and rulings". Since its inception, the AFDO slogan has been "uniformity through cooperation and communication," but the organization's goal has been uniformity without preemption. This view reflects a desire to use the same legal requirements while maintaining authority to enforce local laws.

DEVELOPMENTS BETWEEN 1900 AND 1940

Food Production and Marketing

Prior to the passage of FDCA, Americans generally preserved their own food through the traditional methods of cheesemaking, breadmaking, brewing, fomenting, pickling, salting, drying, and canning. "More was little needed in the way of packaging and labeling of foods produced by these methods. At that time marketers of food had little to use beyond graphics to differentiate their products. By the 1930s food products were being marketed in branded packaging, including Armour meat products and A&P coffee in bags. In the same period, a number of new grocery items appeared, including bread sliced and wrapped prior to sale and frozen foods. Air conditioning of factories allowed dried products to be prepared, packaged, and distributed, leading to the development of such products as gelatin desserts, which were marketed for the first time.

Up through the 1930s, the primary materials used for packaging were paper, paperboard, or tinplate, although glass was beginning to be an important packaging material. Early in the century, the use of corrugated paper containers to replace the wooden box became widespread.

The designation of first self-serve grocery store has been attributed to Piggly-Wiggly, which opened in Memphis, Tennessee. This store was laid out with four aisles for shoppers to walk through to view and select from its 600 items. By the 1920s, the trend toward combination stores had begun, and grocers began stocking meats and perishables because shoppers preferred to make all their food purchases at a single location. By 1930, the national and regional chains were small stores, selling meats; canned goods; dairy products; bulk and packaged cookies, crackers, and bread; and a limited assortment of produce in season.

Public Health

At the beginning of the twentieth century, The changes were due in part to improvements in sanitation, control of infectious disease, and knowledge of nutrition. It became more widely recognized that the food Americans were eating had an important effect on their health. One result of this awareness was that the Federal government took over control of the safety of the food supply by preventing the interstate transportation of unfit food.

The States, for their part, continued to be responsible for food within their respective borders. During Congressional action on the legislation that subsequently became the 1906 Pure Food and Drugs Act, a witness representing a large food distributing company, who appeared before the House Committee on Interstate and Foreign Commerce to oppose passage of the pending food bill, declared that the food industry of the country rested on fraud and deception. "Make us leave preservatives and colouring matter out of our food," he declared, "and call our products by the right name and you will bankrupt every food industry in the country". Wiley suggested that manufacturers and dealers who would otherwise have made pure and properly branded goods were forced by unfair competition to practice the arts of adulteration and misbranding.

During the same period, work of U.S. Public Health Service scientists an the dietary cause of pellagra—a disease resulting from a deficiency of niacin—brought into sharp focus the public health importance of good nutrition and added a new responsibility to the mission of public health officials, which earlier had been limited to sanitation and adulteration. The first 40 years of this century constituted the era of discovery of the nutrition deficiency diseases and isolation of the responsible nutrients.

It suggested that the problems that still remained on the debit side were polio, encephalitis, influenza, and cancer. On the credit side, it listed smallpox vaccination; antitoxin, toxin-antitoxin, and one-dose toxoid for diphtheria; typhoid and yellow fever vaccines; antilockjaw serum; vitamins for scurvy, rickets, and pellagra; and sanitary knowledge to keep foods and water supplies germ free. The author reported on a series of cases of "disease of bad plumbing" that occurred in Chicago in the early 1930s in which food handlers were contaminating food as a result of the use of contaminated water.

Food Regulation

In the early 1900s, the food industry strongly supported national food legislation in order to obtain national uniformity in regulatory requirements to build credibility for the food supply. After considerable debate on the constitutionality question surrounding States' rights, Congress enacted the Pure Food and Drugs Act of 1906. This law was the first national effort to regulate food and drugs by prohibiting the adulteration or misbranding of these products. The Act defined food as including wall articles used for food, drink, confectionery, or condiment by man or other animals, whether simple, mixed, or compound."

One of the goals of the law was to establish national uniformity to reduce confusion in the marketplace. Consistent with the food industry's support for passage of the legislation, the House report stated that the laws and regulations of the different States are diverse, confusing, and often contradictory. What one State now requires the adjoining State may forbid. Our food products are not raised principally in the States of their consumption.

State boundary lines are unknown in our commerce, except by reason of local regulation and laws, such as State pure-food laws. It is desirable, as far as possible, that the commerce between the States be unhindered, one of the hoped-for good results of a national law on the subject of pure foods is the bringing about of a uniformity of laws and regulations on the part of the States within their own several borders.

In spite of these intentions, passage of the 1906 Act did not secure national uniformity in food regulation. The Act did not establish the basis for a comprehensive national regulatory scheme bemuse it applied only to regulation of foods in unbroken packages in interstate commerce—and then only to the actual label of the product. Once the food package was broken, and many packages continued to be in large bulk containers, the States had authority over the sale of the product. In a number of cases, the courts upheld State regulations that imposed requirements different from or in addition to those imposed by the Federal government.

By contrast, early statutes to regulate meat and poultry products dealt with the issue of uniformity. The 1907 Meat Inspection Act required postmortem inspection of all animals and meat prepared for human consumption and transported in interstate commerce. The legislation also gave USDA the authority to supervise both processing and labeling of all meat products, and preempted State requirements for those products when moved in interstate commerce. Prior label approval was required for the marketing of meat products. In the mid-1920s, USDA established a voluntary poultry inspection programme.

The 1921 Annual Report of the USDA Bureau of Chemistry pointed out that both officials and manufacturers complained of the lack of uniformity in the exercise of food control by Federal and State governments:

> Lack of uniformity increases the costs of doing business, and the increased cost is usually passed on to the consumer. It arises not merely from differences in the various laws but also from differences in the interpretation of the laws by the officials in the application by them of different standards to the same product in different jurisdictions.

Concerns about the problem of nonuniform food requirements have, however, persisted to the present, as indicated by an average of one speech on the subject given annually at the AFDO conference.

During consideration of the legislation that was to become FDCA, a 1935 Senate report, in recognizing the "problem of uniformity," noted that the States had unanimously urged the Federal government to assume leadership in modernizing existing law. FDCA continued the authority of Federal officials over foods that traveled in interstate commerce. The Act defined the term *food* in Section;

1. Articles used for food or drink for man or other animals,
2. Chewing gum,
3. Articles used for components of any such article.

In structure, FDCA was a series of definitions elaborating the two basic concepts, adulteration and misbranding. Under the statute, FDA was empowered to regulate all labeling of food shipped in interstate commerce and deal with other matters relating to the safety and wholesomeness of food. FDCA also contained the enforcement remedies available to FDA. The sanctions it authorized include criminal prosecution of individuals and firms responsible for prohibited acts, injunction against such acts, and seizure of adulterated or misbranded goods; the latter sanction is the one most commonly used. FDA has also used several informal remedies such as publicity and regulatory letters, under the provisions of the Act, and recalls through negotiations with industry, which are not explicitly provided for in the Act.

DEVELOPMENTS BETWEEN 1940 AND 1970

Food Production and Marketing

With the advent of World War II came the compelling need for mass production and transportation of food to the troops. In the 1940s, mass production allowed the movement of large quantities of raw materials through production plants, and improvements in conveying led to the use of automation to facilitate production and packaging. Along with the increased need for foods for the military came the development of many substitute foods. At the same time, frozen foods were introduced on a large scale, and vending machines became a new means to sell foods. Supermarkets quickly adjusted to post-World War 11 prosperity, each decade, store inventories grew by 2,000 to 3,000 additional items.

By the 1950s, introduction of the freestanding home freezer provided convenient storage for consumers who selected the newly developed frozen dinners and frozen, ready-to-eat bakery goods. Foreign foods became widely accepted. Increased prosperity led to targeted markets and allowed the food industry to introduce a large number of new products. One such product was instant milk powder, which led to dried milk powders, whey products, cocoa-and strawberry-milk beverages, and spray-dried coffee and tea. Late in the decade, the perceived need for food in bomb shelters resulted in technological advances that spurred the production of new foods with a long shelf-life.

Advances in equipment technology led to a high-quality freeze-dried coffee that quickly became popular. New ingredients, such as off blend and falvouring agents, were also entering the market at this time. Foam-mat drying and related techniques improved the taste of dried milk. Enzyme technology began to be applied in food processing to develop unique products.

Health and organic foods were becoming regular items on the grocery shelf. This decade also saw the introduction of membrane-processing systems, which changed the characteristics of food fluids; commercial use of these systems focused on industry by-products. Extrusion of carbohydrate and protein foods was another innovation that led to a new generation of precooked, ready-to-eat cereals, snack foods, candy bar fillings, breading for fish sticks (that did not require baking), and dried vegetable and animal protein products.

A major advance for the food industry came with the more widespread use of aseptic processing. It is defined as separate high-temperature, short-time sterilization (HTST) of a food product and its packaging material (or container), and the filling of the product in a sterile atmosphere. The major advantages of such processing are improved shelf life, food quality, and nutrient retention; reduced energy use in processing and distribution; reduced storage required for packaging materials; the ability to combine paper, plastic, and metal foils in packaging; and potential expansion of sales into new markets. Which is constructed of a combination of polyester, aluminum foil, and a heat-sealing polyolefin. Although applications of this kind of packaging in the United States, which were intended to allow thermal processing and shelf-stable distribution, have been limited to date, the pouch offers the advantages of light weight, compact size, easy disposal, convenient reheating, and energy savings in terms of processing and distribution. A significant disadvantage is the lack of a practical method for recycling such materials. More recent canning developments have included the two-piece can using draw-redraw techniques, welded cans, and lighter-weight tinplate.

As a packaging material, plastics were virtually unknown before the 1930s. The discovery of polyethylene and the applications developed for polyvinyl chloride during that decade marked the beginning of a revolution in food packaging. The need for and resulting shortage of all materials created by World War II led to major advances in the development of plastics and their applications. Plastic packaging offered the advantages of being lightweight, durable, and unbreakable. Added to the use attributes of plastic were the engineering advantages of mechanical performance equal to metal, ease of design, flexibility, endless moulding and functional possibilities, dispensing capabilities, barrier properties, and controlled permeation. Although many plastics can be recycled, workable methods to sort and use recycling products are still in their infancy.

Several other advances have occurred that have affected food packaging. During the 1950s and 1960s, improved efficiencies in production, light weight packaging, and downsizing led to a decrease in the cost of packaging relative to the cost of food expenditures. Such improvements as lighter-weight steel, the addition of ribs to add strength, better glass distribution, the introduction of polymers, lightweighting of papers, and improved barrier properties led to less material required per package. In addition, packaging assembly lines could operate at greater efficiencies and speeds. More recently, computer applications have improved efficiency and reduced cost by influencing design and manufacturing.

Public Health

Beginning in about 1940, the focus in nutrition research shifted to the determination of human nutritional requirements and the nutritional quality of foods. Starting with the enrichment of flour and bread, fortification of food was implemented as a public health measure to increase the intake of nutrients, and this practice had a dramatic effect on the nutritional status of Americans. In the 1960s, concerns about undernutrition in Americans living in isolated areas of the country came to national attention.

The increase in knowledge of the determinants of growth and development of children led to recognition of the importance of nutrition to maternal and child health. The fact that mortality rates from many infectious diseases such as dysentery and measles were higher in malnourished children than in those who were well nourished led to numerous studies of the interaction of nutrition and infection. Such research showed that malnutrition results in increased mortality from infection, which in turn puts an increased strain on an individual's nutritional reserve.

One of the purposes for passage of FDCA in 1938 had been to address problems associated with misbranding; nevertheless, many of those issues persisted. The violations reported were labeling concerns (many of which are related to the six provisions involved in this study) that continued to occur decades after passage of FDCA. Similar information on more recent cases was not readily available to determine whether these problems have continued to persist.

Food Regulation

The food provisions of FDCA have been amended on a number of occasions, primarily with regard to safety issues. Enforcement strategies for food safety have been built on the premarket approval concept imposed on drugs in the 1938 Act; these strategies place the burden of proof of safety on the manufacturers of new food and colour additives. Statutes that modified FDCA include the Pesticides Amendment of 1954, the Food Additives Amendment of 1958, the Colour Additive Amendments of 1960, and the Animal Drugs Amendments of 1968.

Prior to passage of NLEA, Congress expressly provided for preemption of State food labeling regulation under the Fair Packaging and Labeling Act of 1966. FPLA required that the net weight of a food product, as well as other required information, be accurately stated in a uniform location on the label to facilitate value comparisons. FPLA specifically declared that it is the express intent of Congress to supersede any and all laws of the States or political subdivisions thereof insofar as they may now or hereafter provide for the labeling of the net quantity of contents of the package of any consumer commodity covered by FPLA which are less stringent than or require information different from the requirements of Section of the Act or regulation promulgated pursuant thereto. This Act also authorized FDA to adopt regulations to prevent the nonfunctional slack fill of packages containing foods. To date, however, FDA has not proposed or promulgated regulations to implement FPLA provisions.

Congress amended FDCA in the Regulatory Amendments of 1948, expanding its jurisdiction to include any action with respect to a food that results in the article becoming adulterated or misbranded after shipment in interstate commerce. However, there was no attention to national uniformity in the 1938 Act or the 1948 amendment.

Poultry remained subject to FDA regulation until 1957. Passage of the Poultry Product Inspection Act in 1957 provided USDA with statutory authority for mandatory post-mortem inspection of every carcass and ante-mortem inspection of poultry in interstate commerce, with the Federal government required to pay the cost of the inspection programme. The Processed Products Inspection Improvement Act of 1986 made a basic alteration in the law, to permit less than continuous inspection of processed products; it did not, however, alter the requirement for prior label approval.

The interest focused on the quality of State inspection of meat and poultry slaughter and processing plants engaged only in intrastate commerce. In that year, the Wholesome Meat Act and the Wholesome Poultry Products Act changed the rather casual relationship that had previously existed between the Federal and State levels of regulatory activities. These two laws gave USDA explicit statutory authority to preempt State regulation of meat and poultry products with regard to inspection and labeling for products in intrastate as well as interstate commerce. The States were allowed to maintain their own meat and poultry inspection programmes, provided that within 2 years, each State programme was certified as at least meeting Federal standards. For States that conducted their own inspection programmes, the Federal government was to provide 50 percent of the funding; where no State programme existed, Federal inspection would be provided at no cost to the State. As such, the legislation was a compromise to allow State Directors of Agriculture to continue to operate their

States' programmes and administer the Federal meat inspection programmes. All of these products were to be inspected and labels were to be approved prior to marketing. This requirement applied to all products that contained 2 percent or more of meat or poultry by weight.

DEVELOPMENTS BETWEEN 1970 AND 1990

Food Production and Marketing

In the past decade, aseptic processing has been widely used in Japan and Europe, with growing pressure to adopt this technology, commonly known as refrigerated prepared foods, in the United States. The 1980s also saw increasing interest among consumers in product quality, which has led to a new generation of upscale foods, such as gourmet products and frozen entrees. Wider use of food irradiation has been approved, although it is not yet generally applied to retail products. New nonorganic-solvent techniques, including natural substances (such as water and vegetable oils) and supercritical carbon dioxide, have been developed for decaffeinating coffee and tea products. The increased use of microwave cooking in the home has led the food industry to respond by designing new products and adapting existing ones to meet the demand for products cooked or merely heated by this method. Ten percent of U.S. homes had microwave ovens compared with 75 percent ; by the year 2000, they are expected to be in 90 percent of American homes.

The major growth seen in the use of plastics in food packaging has resulted in potentially thousands of combinations of plastic components currently in the development stage, which offer a multitude of packaging options for the future. Environmental concerns, however, may have an impact on these developments, producing combinations that are more ''environmentally friendly.''

The most recent trend affecting food packaging and the industry has been the development of the category of freshly prepared and catered products. Systems that produce these products utilize controlled or modified atmospheres of oxygen or carbon dioxide in packaging to control microbiological growth. Thus, the shelf-life of these products in a well-controlled distribution system can be extended for a number of days or even weeks. Increasing use of products in this category, however, will require rethinking of the current national food distribution system of centralized, high-volume production facilities, because these products require controlled storage and transportation conditions. Some estimates predict that more than half of all foods consumed will fall into this category.

Today the supermarket is a relatively impersonal, streamlined, one-stop entity designed to maximize efficiency and minimize consumer time for food shopping. These operations carry from 9,000 to 12,000 items in an average of 22,000 square feet of floor space. At the same time, gourmet stores are becoming more prevalent; these operations provide unique food items and more personal service, reflecting a return to the specialty store concept of yesteryear. The number of new products seeking profitable marketing niches continues to explode. Until 1981, an average of 2,500 grocery products was introduced annually. Throughout the 1980s, this number grew steadily, reaching a level of more than 13,000 new products introduced in 1990. With this explosion of new products, the length of time now used to judge success or failure on the shelf has been shortened. In many markets, a 3-month time frame for judging the initial success or failure of a product is common.

Perhaps one of the most interesting shifts in food distribution over the past 15 years has been in the way products are priced and promoted. As a consequence of this shift, manufacturers no longer have complete control over the geographic distribution of products; instead, control has moved from manufacturers to the product distributors and retailers. Trade promotions are conducted regionally, and local price discounts that are targeted to a specific area or company are often, but not always, accompanied by performance requirements in the form of minimum quantities to be purchased or special in-store displays.

Although these local and regional price promotions are developed for good reasons, they also have unintended consequences, including the creation of an active diverting network. As distributors see large regional price differentials, they develop networks for moving products from one region to another. Distributors simply buy in excess of their local needs and move the product to their stores in another region where the discount off-list is not available, or sell to diverters who market the goods to the highest bidder. Although by its very nature this market is impossible to measure, estimates of its volume now range to as much as $10 billion yearly, and it is still growing.

Once a product moves into a diverter network, it becomes impossible for the manufacturer to track, let alone control, its distribution. Products, and therefore labels, designed for one market area will almost inevitably be distributed throughout the United States. Manufacturers faced with inconsistent or conflicting local regulations therefore feel increasing pressure to satisfy several jurisdictions with the same label or design labels for the most restrictive markets. In some cases, meeting all the requirements of all the localities in which the product might find its way to market is impossible. In all cases, the added uncertainty imposes a cost on the industry and the final consumer. The costs incurred in meeting unique requirements need to be weighed in terms of the burden they create compared with the desired outcome. The recurring theme concerns which level of government should be charged with the regulation of food products to ensure that they are properly labeled and reach the market in the most expeditious manner.

Public Health

Concern about the persistence of undernutrition continued in the United States into the early, but it was accompanied by a growing awareness of the problems of excess consumption. Various public and private organizations began to make dietary recommendations on total fat, saturated fat, cholesterol, and sodium as risk factors for heart disease or cancer. Diet and health relationships were the focus of considerable research and debate. Although considerable knowledge has yet to be ascertained about certain dietary constituents and their relationship to chronic disease, labeling concerns have intensified as more has been learned regarding the long-term public health significance of certain nutrients.

Criticism of the information on food labels escalated, spurred by two developments. First, scientific investigations had convincingly demonstrated important linkages between dietary habits and the prevalence of chronic diseases, most notably cardiovascular disease, cancer, stroke, diabetes, and obesity. The American diet was shown to contain considerable amounts of such components as calories, fat, cholesterol, and sodium, which were associated with the incidence of certain chronic diseases. The second development was a response to the first: American consumers became increasingly attentive to choices among foods and sought improved information on the products they were selecting. Food producers and manufacturers responded to this interest by reformulating existing food products, developing new foods, and aggressively marketing those products whose composition could be promoted as reflecting the desirable relationships between nutrition and health. By the late 1980s, however, marketing practices still frequently resulted in incomplete information for making proper food choices. Health and nutrition claims, which were proliferating in the marketplace, were difficult to verify on the basis of the information provided on the existing food label, further highlighting the label's inadequacies.

A number of reports published in the 1980s reviewed the mounting consensus on the relationship between various dietary constituents and chronic disease. Most notable were the *Surgeon General's Report on Nutrition and Health* and the National Research Council report *Diet and Health: Implications for Reducing Chronic Disease Risk*. The scientific evidence in these two documents convinced many policymakers and health professionals that existing food labeling regulations needed to be reexamined to assess whether their provisions afforded consumers adequate information in light of the current scientific consensus on dietary constituents and their relationship to the risk of certain chronic diseases.

Food Regulation

Beyond its recommendations concerning food availability to those in need, the 1969 White House Conference on Food, Nutrition, and Health made a number of recommendations about the provision of information on food packages. Its final report recommended that FDA consider the development of a system for identifying the nutritional qualities of food. The report stated that manufacturers should be encouraged to provide truthful nutrition information about products to enable consumers to follow recommended dietary regimens. Other recommendations included the need to survey various types of consumers to determine their information needs and abilities to use labeling, and develop an educational campaign to teach consumers about how to use food and nutrition information.

FDA promulgated regulations that established the current framework for the nutrition labeling of foods. For most packaged foods, the regulations allowed information on nutrition content to be provided voluntarily but prescribed a standard format. Nutrition labeling was made mandatory, however, on any food to which a nutrient was added or a nutrition claim was made. Subsequently, USDA issued similar guidelines through policy memoranda for nutrition labeling on meat and poultry products. By 1990, more than half of all packaged foods sold in the United States bore some type of nutrition labeling. These changes in food labels that were begun in the 1970s represented a fundamental shift in regulatory philosophy and a major advance in consumer information. From the perspective of the 1990s, however, the adequacy of nutrition information on food labels was questionable, and indeed, during this period, some consumer and professional organizations began to press their concerns regarding nutrition labeling.

Several changes were made to the labeling provisions of FDCA. The amendment to the Health Research and Health Services Act, known as the Proxmire Amendment, was aimed at the labeling regulations for dietary supplements. The Infant Formula Act authorized FDA to adopt regulations requiring that certain nutrient content, labeling, and good manufacturing practices be met in the preparation of these products.

Several Federal agencies decided to review existing food labeling regulations in the United States to determine whether these provisions were still appropriate. FDA, USDA's Food Safety and Quality Service, and the Federal Trade Commission held public meetings on a variety of labeling issues and subsequently published a notice in the *Federal Register* that set out their tentative position on the changes that were needed. The document addressed changes in ingredient labeling, nutrition labeling, label format, open dating, standards of identity, disease prevention claims, imitation and substitute foods, food fortification, and the procedures required for implementing the changes. However, for a number of reasons, including limited scientific consensus and a political climate favouring deregulation, no changes were made at that time. The only regulatory change actually implemented following the late-1970s reform effort concerned sodium content and sodium descriptors. Although several major legislative proposals to reform food labeling were introduced during this period, no bill was enacted by Congress.

An additional concern was the proliferation of health messages appearing in conjunction with the sale of food, which needed more vigorous regulation than was being provided under existing policy. FDA's traditional position on health messages, which had been developed to combat health fraud, was that a food for which a health claim was made was an unapproved drug. This classification required that the product be shown to be both safe and effective for use prior to marketing. By the mid-1980s, however, food manufacturers generally had begun to use nutrient content and disease prevention claims to promote products. In instances in which claims were made without supporting data, the agency could have taken rapid action based on its view that the food was an unapproved new drug. However, the gathering evidence of the relationship between diet and long-term chronic disease led the agency to reconsider its position and announce plans to revise its policy.

Yet in the process of developing a policy to use food packages to provide nutrition information on the role of dietary changes in reducing the risk of chronic disease, FDA was faced with a conflict: the desire to provide more information but not compromise its ability to combat health fraud. The Federal government's slow pace in developing a contemporary policy on health messages because of this conflict—while such messages proliferated in the marketplace—led to the involvement of State Attorneys General in attempts to fill the regulatory gap that they perceived existed in preventing consumer fraud in product labeling and advertising claims. The State Attorneys General pursued a regulatory course parallel to FDA's traditional policy; that is, such claims required that the products be considered unapproved new drugs and removed from the marketplace.

Efforts to reform the current policy on food labeling, especially in regard to nutrition information, were proceeding in several arenas. Improved food label information was more universally viewed as a way to assist consumers in making food choices that would be more healthful. In the spring of 1989, major new legislation was introduced in Congress to mandate nutrition labeling for food products under FDA jurisdiction. By August, FDA, in cooperation with USDA, announced plans to reform nutrition information and other aspects of food labeling. In an Advance Notice of Proposed Rulemaking, the agency outlined a plan to elicit input from interested parties about proposed changes in the nutrition label by requesting written comments, holding public hearings across the country, and contracting for a study to address potential label changes. The U.S. Department of Health and

Human Services (DHHS) and USDA contracted with the Institute of Medicine's Food and Nutrition Board to conduct a study on various aspects of existing nutrition labeling practices.

FDA reproposed rules for allowing health messages to appear in conjunction with the sale of foods. The agency proposed to allow the use of six specific statements of the relationship between diet and disease and establish a mechanism by which other relationships might be approved for label use. FDA proposed regulations for mandatory nutrition labeling of foods under its jurisdiction as the first phase of its food labeling reform initiative. These proposals also focused on reference values, nutrient content, serving size, and cholesterol labeling terminology. Subsequent proposals were to include proposed reform of ingredient labeling, standards of identity, label format, health claims, and label descriptors.

The new law was concerned specifically with FDA's food labeling authority, addressing issues similar to those already under review by the agency. It required mandatory nutrition labeling on most packaged foods and voluntary nutrition information for produce and seafood; specified the nutrient content information that must appear on the label; provided for certain descriptive terms to be defined and claims to be allowed; established a petition mechanism for additional nutrient and health-related claims; required that consumer education be undertaken; provided for State enforcement of Federal requirements in instances in which the Federal government had not acted; revised certain requirements for ingredient listings and standards of identity; provided for national uniform nutrition labeling; and specified the effective dates for implementation of various provisions of the Act. Successful passage of the legislation required the support of industry, which had, from the beginning of the debate, favoured Federal preemption of all State food safety and labeling requirements. As passed, NLEA provided for Federal preemption of State food labeling requirements and specifically exempted State food safety requirements from preemption.

CURRENT SITUATION

Today, meat and poultry inspection programmes are essentially federalized. In the past decade, as a result of technological changes, the amount of meat and poultry inspected has increased markedly while the number of staff has remained essentially the same. States that have continued to operate their own programmes must annually review whether paying half the costs of running a State inspection programme makes sense when the Federal government is willing to take over the programme and pay all of the costs. Many States have opted for Federal inspection for budgetary reasons. In cases in which a State fails to meet Federal requirements, the Federal government assumes responsibility for the inspection of its intrastate plants. Currently, 28 States continue to operate their own meat inspection programmes, and 24 still operate poultry inspection programmes. The current USDA meat and poultry inspection budget is about $500 million, about the same as FDA's entire budget.

State and local regulatory officials have long been concerned about the lack of uniformity among food and drug laws. In response to this concern, AFDO developed and has used as its major uniformity tool the Uniform State Food, Drug, and Cosmetic Bill as a model law for all States. AFDO has continued to amend the model bill to keep it current with changes made to FDCA. One major step to assist that effort was the addition of several provisions providing for automatic adoption of Federal statutes and regulations. The result has been substantial uniformity written into law by State legislatures and implemented by State enforcement officials in their regulations. If enacted by every State, the Uniform Bill is seen by AFDO as providing a sound basis for national uniformity in the regulation of food labeling. Currently, 45 States have adopted the model law; 23 States have adopted all or parts of Federal regulations by reference.

Unfortunately, the adoption of uniform laws, codes, and some implementing regulations, and their interpretation have not resulted in uniform enforcement procedures by all State agencies with jurisdiction over food products. It often remains unclear why State regulation arises. One possible explanation may be that States see practices that FDA could address under the law but does not or at least not to States' satisfaction. In other words, States would favour a different Federal policy. Another possibility is that States see practices that Federal requirements simply do not address or would be unlikely to address because of their local character.

The extent to which State requirements can be categorized as representing distinctive, genuine local needs or reflecting different judgments on how food products ought to be labeled has a bearing on the assessment of the adequacy of Federal implementation. However, such information is not readily available. There are more than 380 State agencies that carry responsibilities similar to those of FDA, which results in understandable differences in regulatory philosophy and enforcement procedures. Since AFDO's membership is drawn from regulatory officials in Federal, State, county, and local governments, the organization has continued to encourage and support uniformity of food laws and regulations among these various jurisdictions.

Current State activities that complement the Federal government's activities in food control are embodied in a number of programmes and services. Most States have differing organizational structures, methods of programme administration, and arrangements with FDA for the various food enforcement activities performed in their jurisdictions. Some States have two departments involved in these activities, but most have three. In the annual report FDA prepares on these State efforts, the food programme is divided into the following 13 areas:

1. Aquatic products
2. Bakeries
3. Candy manufacturers and repackers
4. Canneries
5. Food service

6. Food storage
7. Grade A milk processors
8. Grains
9. Manufactured milk products
10. Retail level establishments
11. Shellfish
12. Soft drink bottlers
13. Other food activities

Reports indicate that in fiscal year 1989, a total of 145.53 million was spent on all these food activities by 46 States and Puerto Rico; this figure represents 74 percent of the total 196.06 million spent for all food and drug control activities by the States. The food categories listed above are not subdivided to permit an assessment of actual food labeling activities performed by State agencies. Provides a review of total food expenditures by programme category.

Federal and State regulators have worked together in a number of ways to promote uniform enforcement procedures. The FDA publication *State Programmes and Service in Food and Drug Control* noted the following initiatives:

- Joint FDA/State inspection of specific establishments or industries to effect both interstate and intrastate correction of violative practices, and for training purposes;
- FDA field office conferences with State counterpart officials to promote mutual understanding and agreement on current consumer protection priorities and planning compliance activities;
- Formal training courses for State and local regulatory officials, held annually across the country in up to 50 locations, which cover food and drug issues to promote state-of-the-art knowledge and uniform inspection/analytic procedures patterned after FDA practices;

- Technical and consultative assistance provided by FDA to States that have primary responsibilities in milk safety, retail food protection, and shellfish sanitation;
- Development of communication networks among the cooperating State agencies and FDA headquarters/field offices;
- Distribution to State officials of various FDA technical and procedural manuals and information;
- Voluntary worksharing agreements under which FDA field offices and State agency managers delineate establishment and/or industry coverage to avoid duplicative or unplanned concurrent inspections;
- Coordinated response plans for emergencies including agreements with cooperating State agencies to facilitate communications and decisionmaking in emergency situations related to food industries or products; and
- The State contract programme, in effect for 14 years, which has had a significant impact on improved uniformity in State inspectional procedures.

TABLE: Total Food Expenditures by Programme Category

Programme Category	Expenditures (millions of dollars)	Percentage of Total	Frequency
Food service	31.20	28.32	38
Grade A milk	21.43	19.45	38
Retail establishments	16.10	14.61	42
Shellfish	7.50	6.91	25
Manufactured milk	6.95	6.31	32
Canneries	5.36	4.86	25
Warehouses	3.58	3.25	40
Aquatic products	3.03	2.75	29
Bakeries	2.92	2.65	40
Bottling plants	1.35	1.23	38
Candy	0.80	0.73	36
Grain	0.69	0.63	17
Miscellaneous	9.27	8.41	33
TOTAL	110.18		

The current regulations governing the labeling of food are located principally in the following sections of the Code of Federal Regulations (CFR):

21 CFR	1.20	Presence of mandatory label information
	1.21	Failure to reveal material facts
21 CFR	101.1	Principal display panel of package form food
	101.2	Information panel of package form food
	101.3	Identity labeling of food in package form
	101.4	Food; designation of ingredients
	101.5	Food; name and place of business of manufacturer, packer, or distributor
	101.15	Food; prominence of required statements
	101.18	Misbranding of food
	101.22	Food; labeling of spices, flavourings, colourings, and chemical preservatives
	101.105	Declaration of net quantity of contents when exempt
21 CFR	102	Common or usual name of nonstandardized foods
21 CFR	103.14	General statements of substandard quality and sub-standard fill of container

Implications

Throughout most of the twentieth century, the increasing amount of food products in interstate commerce has led to persistent problems for State regulators. More recently, the increasing complexity that exists at all levels of government has expanded the areas of overlap between Federal and State activities. In some cases, the enactment of local statutes has further added to the practical difficulties of determining the actual or appropriate relationships among agencies. For decades, obsolete State provisions have not been repealed, despite enactment of the AFDO Uniform Bill. In addition, individual States have frequently modified the Uniform Bill prior to enactment or instituted their own interpretation of similar language, thus creating local exemptions to the general rule of uniformity. Finally, a number of State agencies have adopted additional regulatory requirements on an ad hoc basis to protect local interests (e.g., for indigenous agricultural products or to obtain local political support). These diverse sources of State requirements have been viewed by industry to interfere with interstate commerce and create a seeming disparity between Federal and State laws and regulations. As a result, food manufacturers that wish to market a new food product complain that they must attempt to achieve compliance with Federal requirements and then wait to see if any State challenges the product for noncompliance with its requirements.

The dramatic changes and advances over the past 50 years in the manufacturing, packaging, and marketing of foods have presented regulatory concerns for FDA and the States. For example, texturized vegetable protein products that were introduced in the 1970s mimicked tuna, chicken, pepperoni, and cheddar cheese in their appearance, taste, smell, and mouth feel. Federal regulators were justifiably concerned about the potential labeling of these products under FDCA Sections with the potential for nomenclature issues that these developments might represent. In response to new technologies, common or usual names were established for onion rings made from dried onions, potato chips from dried potatoes, and fish sticks or portions from minced fish.

The use of new and innovative packaging materials has increased the potential for the marketing of misleading containers under FDCA Section and abuse of the prominence of information provisions under FDCA Section. Similarly, the use of artificial flavours and colours and chemical preservatives in the preparation of new and unique substitute foods, specialty foods, and snack foods has emphasized the importance of the Section labeling requirements. Hutt and Merrill have pointed out that in recent years the ingredient statement has allowed consumers to identify and avoid specific ingredients, in contrast to the view in 1938 that ingredient labeling was strictly an economic issue. The very fact that the supermarket of today carries from 9,000 to 12,000 items and a new item can be introduced and withdrawn in a period as short as 3 months highlights the importance of minimizing the burden of all food labeling requirements.

For many years, Federal, State, and local food regulators have supported the goal of uniformity through means other than preemption. Although much has been accomplished, many regulators and the food industry remain concerned that the goal has not been achieved to the extent desired. For those holding this view, NLEA is seen as an opportunity to accelerate the process of achieving national uniformity in food labeling while continuing to provide for joint Federal/State initiatives and a meaningful role for States in food labeling policy development and implementation.

4

The Nature of Food Hazards

CAUSE FOR INCREASING CONCERN

There have been dramatic changes in the Indian food supply. These changes have contributed to recent outbreaks of infectious foodborne illness, which in turn led to the request for this committee to examine aspects of the Indian food safety system. The committee recognizes the growing concern for controlling the microbiological hazards related to food, but believes that government attention must also be addressed to chemical and physical hazards. This chapter is organized in two parts: the first describes major changes that affect the epidemiology of infectious foodborne disease, and the second describes examples of potential chemical hazards which have emerged in part from some of the same changes in the food supply.

THE EPIDEMIOLOGY OF FOODBORNE DISEASE

Recent outbreaks of foodborne disease caused by many different pathogens and involving a variety of food products have been the subject of headlines, but it is unclear whether the incidence of foodborne disease has increased over the last generation. The major reason for the uncertainty is that the lack of a national foodborne-disease surveillance system has prevented the study of trends in disease rates. What is clear is that factors affecting the potential safety of the nation's food supply have changed dramatically over the last generation and justify concern that the incidence of foodborne disease is high and may be increasing.

At least five trends contribute to the possible increase in foodborne disease: changes in diet, the increasing use of commercial food services, new methods of producing and distributing food, new or re-emerging infectious foodborne agents, and the growing number of people at high risk for severe or fatal foodborne diseases.

Diet

Annual food expenditures in the United States, as a share of disposable personal income, decreased from 14 percent to 11 percent from 1970 to 1996. No industrialized nation spends a smaller share of its wealth on food than the United States. For much of the population, readily available food is more varied and more affordable than ever before. For example, in the 1960s, an average US grocery store had fewer than 7,000 food items available. Today, an average US grocery store sells about 30,000 food items, and over 12,000 new products are introduced each year.

During this time when relative costs of the US food supply are decreasing, per capita consumption of many foods has changed substantially. Public health efforts to promote a "heart-healthy" diet have helped to boost the consumption of fresh fruits and vegetables. On a per capita basis, in 1995, Americans ate about 31 lb more commercially grown vegetables, including potatoes and sweet potatoes, and 24 lb more fresh fruit than in 1970. As the consumption and variety of produce have increased, so has the importation of produce from developing countries. The General Accounting Office estimates that in 1995 one-third of all fresh fruit consumed in the United States was imported. Food imports have increased both because of lower production costs in foreign countries and because of consumer demand for year-round supplies of fruits and vegetables that have limited growing seasons in the United States.

For example, 17 percent of cantaloupes, 52 percent of green onions, 36 percent of cucumbers, and 34 percent of tomatoes sold in the United States in 1996 were grown in Mexico. Seasonally, as much as 79 percent of a particular commodity consumed in this country has been raised in Mexico alone, and the percentage of produce from other developing countries consumed here is growing rapidly. Fresh produce items were the leading vehicle associated with foodborne disease outbreaks in Minnesota from 1990 to 1996, accounting for almost one-third of all outbreaks. This percentage is higher than that available from national foodborne disease surveillance data and possibly reflects more active surveillance in that state.

Other trends in the United States are the decreasing consumption of beef and the increased consumption of chicken and seafood. In 1970, the average American consumed 79 lb of beef, 27 lb of chicken, and 12 lb of fish and shellfish; in 1996, annual per capita consumptions were 64, 50 and 15 lb, respectively. Contamination of red meat with *Salmonella and Escherichia coli* 0157:H7 remains important, but the risk posed by chicken as a vehicle for *Campylobacter and Salmonella* has grown substantially.

Cultural changes affect not only what Americans eat, but also where they eat and how their food is prepared. Increasingly, Americans have time-pressured lifestyles. Saving time and effort in shopping for and preparing food will continue to be important for many Americans. Households with a single parent or two working adults often face particular pressures in food shopping and meal preparation that can affect food selection and safety.

With less time spent in the kitchen and greater availability of high-quality, ready-to-eat dishes and convenience items, Americans' food preparation skills are diminishing. As a result, appreciation of simple but critical food safety techniques, such as washing hands and utensils and storing foods at optimal temperatures, has likely diminished.

Individual food tastes and preparation styles are brought to the United States from around the world, and the increasing ethnic diversity of the American population may affect food safety in several ways. Different ethnic groups have different concerns and practices regarding food safety, and this could affect the activities of immigrants as food preparers at home and in the workplace. New food risks can arise as immigrant populations adapt traditional preferences and practices to their new environment. Food safety education programmes should consider the food beliefs and practices of various cultural groups.

Although data show a rise in perceived risk of foodborne illness among consumers, attitudes do not always translate into improved food-handling practices. Over half of all shoppers report washing their hands and/or food preparation surfaces, yet only 28 percent know that cooking temperatures are critical and that foods should be refrigerated promptly. A recent Food and Drug Administration study found that many US consumers still eat undercooked hamburger meat and raw eggs.

Commercial Food Services

The percentage of the food dollar spent on food consumed away from home has risen dramatically over the last three decades. In 1970, only 34 percent of our food dollars were spent eating away from home. In contrast, in 1996, 46 percent of our food dollars were spent for meals and snacks prepared outside the home. Institutional feeding sites serve a wide range of people—from very young children in child care centres to the elderly—in congregate sites for meals, alternative-care centres, and nursing homes. Meals are also prepared as takeout foods from supermarkets and convenience stores; many of these meals include one or more cold food items, such as delicatessen sandwiches and salads that require extensive food handling and are not cooked before consumption. These changes have led to an increase in the number of people handling food and the potential for an increase in the transmission of foodborne diseases from food handlers to consumers.

The average food handler in this country earns the minimum wage, lacks sick leave and other health benefits, and has very limited opportunities for advancement. These jobs are filled by people with few employment opportunities and low economic status, conditions that may be related to a high incidence of intestinal diseases and low rates of routine hand washing. Recent evidence from Minnesota demonstrates the increased risk of food handler associated transmission of *Salmonella typhimurium;* this finding was documented because *Salmonella* isolates in Minnesota undergo molecular characterization and epidemiologic investigation. Food handlers in other areas of the United States probably play a similar role in the transmission of *Salmonella* and other enteric agents.

Production and Distribution

The changing availability and sources of our food supply have brought changes in methods of producing and distributing food. Today, large manufacturing plants can process quantities and types of products that two decades ago would have required many smaller plants. Although the ability to control possible hazards increases when there are fewer plants, the potential for larger outbreaks—even if product contamination is minimal—is evident. Thousands of people become ill during such outbreaks. For example, contamination of tanker trucks with nonpasteurized liquid eggs and subsequent use of those trucks to haul pasteurized ice cream mix most likely led to the largest documented outbreak of salmonellosis in the United States. That outbreak showed that even sporadic low level contamination of a single product can result in a major epidemic of foodborne illness because of the quantity of product consumed.

In the spring of 1998, a cereal produced by the largest generic label manufacturer of cereal in the United States caused a national outbreak of *Salmonella agona* infection. Outbreaks such as this can be difficult to detect because contamination of the product is sporadic and the product is marketed widely.

The driving forces of globalization, advanced technology, and economic competitiveness have dramatically affected the structure and practices of livestock and poultry production. Herds, flocks, and other populations of food animals, including fish and seafood, are increasingly concentrated in fewer and larger production units. The traditional farmstead model of the past, often characterized by multiple species and small numbers of food animals reared on a single farm, has been replaced by specialized, large-scale production systems. For example, from 1994 to 1995, the number of US hog operations decreased from 207,980 to 182,700, but both the inventory of hogs and the total number of hogs produced increased during the same time. That mirrors the general agribusiness trend, and the restructuring continues.

Infectious Foodborne Agents

The role of new or re-emerging causes of foodborne diseases is well-recognized, but the size of the increase in risk is unknown. Emerging and reemerging infections have been defined as new, recurring, or drug-resistant infections whose incidence in humans has increased in the last two decades or whose incidence threatens to increase in the near future. Recent examples of newly recognized agents are *E. coli* 0157:H7, other pathogenic *E. coli, Cyclospora, and Cryptosporidium.* Old agents re-emerging in new vehicles or product streams include *Salmonella enteritidis* in eggs and hepatitis A virus in produce. Finally, there have been recent increases in antimicrobial resistance of pathogens, such as *S. typhimurium* DT104 and *Campylobacter jejuni* in humans, which have been attributed to the use of antimicrobial agents in food animals.

The frequently cited annual estimates of foodborne disease (up to 81 million cases) and 9,000 associated deaths are based on assumptions that do not necessarily reflect the current national foodborne disease problem. Those estimates must be qualified for two reasons. First, no comprehensive population-based studies of gastrointestinal illness in the community have attempted to determine what proportion of these illnesses is due to consumption of contaminated food and what proportion is from other sources. Second, foodborne illness can cause clinical conditions not characterized by gastrointestinal symptoms, such as congenital toxoplasmosis, hemolytic uremic syndrome, salmonella-associated septicemia, and invasive *Listeria* infections. In the absence of comprehensive estimates of foodborne disease incidence, the only data available are those from the FoodNet programme.

Data from FoodNet-an effort sponsored by the Centres for Disease Control and Prevention, the US Department of Agriculture, FDA, and the health departments of California, Connecticut, Georgia, Minnesota, and Oregon-have demonstrated that the incidence of diarrheal illness in the United States is about 1.4 episodes per person per year, or some 370 million episodes of diarrhea each year. Population-based studies of diarrheal disease conducted from 1948 through 1971 in Ohio and Michigan demonstrate an incidence of about one episode per person per year. The reason for the apparent 40 percent difference between these estimates is unclear, but it is evident that the problem is large. If only 25 percent of diarrhoeal disease is food-related, the burden of foodborne diarrheal disease in the United States far exceeds current estimates.

Fatal Foodborne Disease

The risk of foodborne disease is related to several factors, including the presence and dose of a pathogen or toxin in food, the virulence of the pathogen or toxin, the mechanisms of transmission, and the susceptibility of a host. Many factors influence susceptibility to infection and the severity of disease, including age, the use of immunosuppressive agents, and disease states that increase immuno-suppression.

Young children are more likely than adults to develop illness from selected pathogens, and children have been high-profile victims of several foodborne disease outbreaks in recent years. Young children today are likely to be exposed to a broader range of foodborne diseases than was true a generation ago, because families eat out or take prepared food home more often and children receive more community and out-of-home care at younger ages.

Elderly persons are particularly susceptible to illness from foodborne disease, and the number of the elderly in the United States is increasing rapidly. From 1965 to 1995, the number of Americans aged 65 years or older grew by 82 percent. As the 75 million persons in the "baby boomer" generation (equaling one-third of the population) age, the United States can expect one fifth of the population to be over the age of 65 within the next three decades.

As the population ages, the incidence of cancer and other chronic diseases is likely to increase. These diseases will require an increased use of chemotherapeutic regimens, drugs to deal with rejection of solid organ transplantations, and antimicrobial drugs that have important effects on the normal (and beneficial) bacterial flora of the intestinal tract. Together, these factors predispose people to both the occurrence and the serious outcomes of foodborne disease. The increasing number of people with immunosuppressive diseases, such as human immunodeficiency virus, also contributes to the public health importance of food safety.

Chemical Hazards Associated with the Food Supply

Changes in our society and our food supply have raised new concerns about food chemical safety. Some of the changes that have raised new concerns about foodborne infectious disease are also affecting how government agencies carry out the more familiar task of protecting the food supply from toxic chemical agents.

Foods are themselves a complex collection of naturally occurring chemicals that have nutritive, organoleptic, and pharmacological functions and occasionally toxic effects. Naturally occurring toxicants probably present risks second only to those imposed by microorganisms. These natural toxicants include seafood toxins and foodborne mycotoxins.

Other chemicals are introduced into foods intentionally or unintentionally. Intentionally added substances include food and colour additives, flavours, enzymes, vitamins and minerals, and other ingredients that help to add value or characteristic properties or functions to a food. Because of broad public concern about synthetic chemicals, the toxicologic profiles of many of these synthetic components of food are much more complete than those of natural components of food. As a result, the risks posed by regulated food additives are generally better characterized than those of many naturally occurring substances.

Unintentional additives can include environmental or industrial contaminants as well as some substances used in food production but not intended to be part of the food. The migration of food production substances into the final product is generally very low, but must be carefully regulated to ensure safety. Examples include sanitizers used to keep food production surfaces safe, packaging materials used to keep food safe and fresh, pesticides used on crops and drugs used in animals to mitigate damage, disease, microbial toxin production, and general food losses. Environmental or industrial contaminants are not sanctioned but have the potential to enter the food chain. Examples of chemical contamination incidents are methyl mercury in fish, mistaken mixing of polybrominated biphenyls (a fire retardant) into animal feed, and leakage of ammonia refrigerant into frozen foods.

Some people are sensitive or allergic to chemical constituents that are harmless to the rest of the population. These allergic reactions are estimated to occur in 1 to 2 percent of adults and 5 to 8 percent of children. Although serious reactions are rare, it is estimated that several dozen deaths occur each year because of allergic reactions to food. More than 160 foods and food-related substances have been identified as being able to cause allergic reactions. However, in the United States, more than 90 percent of allergic reactions appear to be caused by just eight food types: peanuts, tree nuts (such as walnuts, pecans, almonds, hazel nuts), crustacea (shrimp, lobster, and other shellfish), eggs, milk, soy, fish, and wheat.

Consequently, the subject of chemical hazards in food is complex and includes consideration of potential risks that vary widely in scope and severity. Many chemical hazards associated with foods have been recognized only in the last century as advancements have been made in chemistry, toxicology, and risk assessment. Public concern over these hazards has grown in recent decades, in line with the increasing distrust of chemicals generally and of their use in the environment. Indeed, there have been episodes of chemical intoxication (usually arising out of accident or occupational exposure) with tragic consequences. Although most experts agree that the more serious hazards in the American food supply are not chemical but microbiological, public concern has demanded that proportionally more regulatory resources be applied to chemical hazards.

New Food Components

The American public's growing interest in the relationship between diet and health has led to an increased demand for foods or food constituents that not only have nutritive value but also hold promise for prevention or even treatment of disease. These products have been referred to as dietary supplements, functional foods, pharma-foods or nutraceuticals. In a recent study, more than two-thirds of surveyed households reported use of a vitamin, mineral, or herbal supplement within the previous six-month period. The Dietary Supplement Health and Education Act of 1994 eased restrictions on certain statements of nutrition support made for supplements and exempted them from the safety approval requirement applicable to conventional food additives. That legal change helped spur the growing market for supplements of all types, including herbal products, and raises food safety concerns. Some supplements and herbal products on the market may pose a risk of adverse health effects because they are not required to meet specified safety standards before being sold. They may thus contain varying amounts or unknown or inadequately characterized ingredients that can have pharmacological activity that has not been adequately characterized.

Food processors are examining relatively new sources of ingredients for more conventional functional properties. For example, gums and fibers such as konjac flour, tara gum, inulin, or psyllium fiber may provide bulk and texture to foods, yet have had limited food use in the United States. Potential broad use of these ingredients in domestic foods raises important safety questions: are there significant intestinal effects such as blockage or reduced transit times? Are there impacts on vitamin uptake from the intestinal tract or on osmotic balance? What are the potential allergy risks from increased exposure to plant-sourced materials? These are a few of the safety questions that may be raised when a new ingredient source is considered.

Food processors are also utilizing macronutrient substitutes, such as nonnutritive sweeteners and fat replacers in many food products. These macronutrient substitutes have potential value by lowering calories, sugar, or fat in food products. Because dietary quantities of these substitutes could be substantially higher than the amounts of typical food additives, the assessment of their safety can be particularly important as well as particularly difficult.

New Food Technologies

Modification of plants or animals via genetic engineering can improve yields and increase resistance to pests. This new technology might offer improvements in food safety through increased resistance to molds that produce food mycotoxins or through lower levels of allergenic proteins, fatty acids, or other undesirable components of food. However, there are important differences between countries in how food products from genetically modified organisms are regulated. Several products derived from genetic engineering have been declared safe by US regulators, but many European countries have either forbidden their sale or insisted on what marketers believe is disparaging labeling. Concerns about the safety of products from genetically engineered plants and animals are only partially resolved.

Food irradiation is not a new technology. Irradiation of fruits and vegetables decreases the risk of pathogens and extends shelf life. Public concerns about the safety of irradiated food, fueled in part by a lack of consumer education, limit its use in the United States. These concerns prompted the requirement that certain irradiated foods be labeled as such. The extent to which this labeling requirement has limited subsequent adoption of the process is unknown.

Re-emerging Toxic Agents

As the science of toxicology has progressed over the last 50 years, questions about the safety of chemicals in foods have become more sophisticated. When tougher safety standards are applied to new food chemicals or reapplied to old ones, new issues of toxicity emerge. The cycle of reevaluating safety standards began in the years after World War II when advances in the understanding of the mechanisms of carcinogenesis coincided with the increased use of rodent bioassays. The newly focused attention to carcinogens that ensued led Congress to pass the 1958 "Delaney clause" proscription of carcinogens. The focus on carcinogens in food continues to consume substantial testing and regulatory resources. Further advances in biomedical research have raised new concerns and new standards of safety that now address teratogenicity, reproductive toxicity, mutagenicity, hormonal effects, and immunotoxicity.

The effects of food constituents on hormonal function is of concern as medical research has indicated the importance of certain hormones in regulating diseases such as breast cancer and osteoporosis. This phenomenon, sometimes referred to as endocrine disruption, led Congress in 1996 to mandate special regulatory attention to the issue and inspired the Environmental Protection Agency (EPA) to convene an expert panel, the Endocrine Disrupter Screening and Testing Advisory Committee. That committee has attempted to define endocrine disrupters, determine analytical methods, and recommend how the disrupters should be regulated. Potential candidates for evaluation include constituents of some food packaging materials, pesticides, and natural food constituents, such as genestein in soy.

Some groups have recently asked whether chemical safety assessments provide adequate safety margins to protect children. That concern draws attention because new pesticide legislation directs EPA to consider the aggregate risks posed by any pest control agent; that is, exposures should be assessed for all potential sources of a chemical and for other chemicals with a similar mechanism of action. Another concern is the emergence of evidence of subtle developmental and behavioural effects of relatively low concentrations of chemicals (for example, lead).

In summary, chemical hazards, as well as foodborne pathogens, present new and changing challenges to the food safety system. Federal food safety efforts will need full integration and sufficient support to meet these challenges.

Physical Hazards

The foods we consume begin as raw agricultural commodities grown in open fields or waters or raised in a variety of production facilities, such as barns, coops, pens, and feedlots. Rocks, stones, metal, wood, glass, and other physical objects can become part of raw ingredients. Further contamination can occur in the transport, processing, or distribution of foods because of equipment failure, accident, or negligence.

Foreign physical materials in foods can cause serious harm to consumers. Protective devices that remove or prevent physical hazards include metal detectors, magnets, sieves, traps, scalpers, and screens. Other effective means of protection are production plant policies against the use of glass, wood, or nonferrous metal where possible; employee training; quality audits of ingredient suppliers; and sensory tests.

Federal agencies have established "defect action levels" for natural or unavoidable defects that do not affect human health, such as stems or pits in fruits and vegetables, bone in mechanically deboned meats, and microscopic insect fragments, which are primarily of aesthetic concern and do not present a safety hazard.

The Changing Nature of Food Hazards

- Changes in the risk of infectious foodborne disease are due primarily to:
- Changes in diet,
- Increasing use of commercial food service and in food eaten or prepared away from home,
- New methods of producing and distributing food,
- New or re-emerging infectious foodborne agents, and
- The growing number of people at high risk for foodborne illnesses due to:
  - Increasing number of elderly, and
  - Increasing number of people with depressed immunity or resistance to infection.
- Changes in chemical hazards associated with the food supply must be monitored and evaluated; these include:
- Increased use of dietary supplements and herbal products without requirements to meet specified safety standards;
- New food components that mimic attributes of traditional food components;
- Introduction of new food technologies and processes;
- Changes in presence of food toxins and additives, including unintentional food additives; and
- Presence of physical hazards associated with new technologies or sources of foods.

5

Effective Food Safety System

THE CONSTITUENTS

The definition, historical significance, and current status of food safety in India as well as recent trends in and reasons for concern about foodborne hazards. This chapter turns to the features of an ideal food safety system and to a somewhat detailed description of its attributes. Constructing a model system establishes a benchmark that provides both a point of reference for judging actual systems and a goal to be achieved.

THE MISSION OF THE SYSTEM

There is universal agreement on the need for safe food, but no consensus on the means by which safe food is secured. The effectiveness of a food safety system begins with a clear, unified mission that focuses and integrates the varied needs and responsibilities of all stakeholders, gives the stakeholders a basis for achieving the goals of the system, and is broadly accepted.

The committee defines safe food as food that is wholesome, that does not exceed an acceptable level of risk associated with pathogenic organisms or chemical and physical hazards, and whose supply is the result of the combined activities of Congress, regulatory agencies, multiple industries, universities, private organizations, and consumers. The mission of a food safety system should be stated as an operational charge that uses and reflects that definition. The committee defines the mission as follows:

The mission of an effective food safety system is to protect and improve the public health by ensuring that foods meet science-based safety standards through the integrated activities of the public and private sectors.

The remainder of this chapter focuses on these matters. Activities that are implied in the definition and are incorporated in achieving the mission to ensure safe food include: adequate monitoring and surveillance; science-based research and development; incorporation of the tenets of risk analysis, including risk assessment, risk management, and risk communication; good practices in food production, processing, manufacturing, retail sale, transportation, preparation, and handling; and appropriate technical assistance and education.

General Attributes of the System

An effective food safety system is an interdependent system composed of government agencies at all levels, businesses and other private organizations, consumers, and supporting players. The system is dynamic and aligned to the unified mission of improving food safety so as to maintain and improve the public's health and well-being.

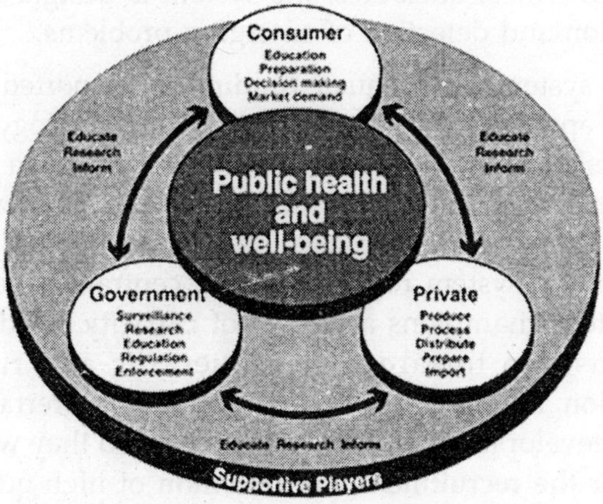

Fig. Attributes of an Effective Food Safety System

Figure depicts the interrelationships of an effective food safety system. Although the players have key, independent functions, they must implement many of their actions through strong partnerships. The system is built on the flexibility and adaptability of the players and on the nature and course of their relationships, and these affect their ability to prevent, identify, and resolve food safety issues in the most efficient, effective, and cost-beneficial manner.

A Dynamic Interdependence. Partners in the system include government, private industry, and consumers. Supportive players, who are critical to the integration of the attributes of research, education, and information, include universities and colleges, the news media, and focused special interest organizations, among others.

The system is responsive to the expedient issues of today but also evolves strategically to meet future challenges. It must be effective in both the domestic environment and the new global food environment, with increasing international trade. The system is science-based with strong emphasis on risk analysis and the use of data. It is attentive to learning through the use of feedback loops and continuous improvement. Although responsiveness to and coordination of food safety crises are critical attributes, the system is designed to stress prevention and detection of emerging problems.

The system has adequate funding, is supported by strong research and education components, uses technology adequate to the task, and is integrated to achieve its mission. Statutory and regulatory authority promote the system's horizontal and vertical integration. Because integration is such a critical attribute, the system requires strong, centralized leadership. The system champions a culture of capacity building. With the transition toward a new scientific and risk-based foundation, agencies will encourage and fund retraining and further development of their employees, and they will initiate plans for the recruitment and retention of high-quality staff with the skills and knowledge to enhance the new system and its changing operations and focus. Capacity must also include consumer knowledge and practices, taking into account cultural sensitivities and practices.

An effective system is commensurate with today's driving forces, trends, and societal expectations. Partnerships will expand with the growing recognition that government cannot abandon our food safety problems to private industry and consumers.

The effective system stresses the inclusion of its players in their roles but also acknowledges the need for effective regulation, compliance, and enforcement. Although some of the roles will need to change, the system functions in an environment of trust and respect. The globalization of the food system and other factors that can increase the risk of foodborne incidents mandate urgency in adopting the system.

The effective food safety system is focused on public health, and its many actions are aligned to achieve a safer food supply, improve public health, and instill consumers' confidence in both the system and their role in improving it. Finally, the dynamic interconnectiveness also promotes the attainment by all of the players of both responsibility and accountability for making the food safety system perform optimally. The following sections describe in more detail the attributes of an effective food safety system.

The Importance of Partnering

Food safety is the responsibility of numerous and diverse stakeholders, and partnerships provide the links that are necessary to build a coordinated and cohesive framework for action. Partnerships can improve efficiency and provide a mechanism for information and technology transfer. Interaction and communication through partnerships lead to cooperation and collaboration among public and private interests. Partnerships can also help to integrate regulated activities with important non-regulatory components of the system.

Incentives can greatly influence and facilitate the building of effective partnerships. "Positive" incentives are often financial. "Negative" incentives can include the desire to avoid legal or regulatory action or media attention. Another incentive for partnerships can be the generation of new and useful information that improves production and processing capabilities simultaneously with improved control of risk.

The market is an important incentive for private industry. As global markets and domestic consumers expect safer food, safety itself can become a factor in differentiating products. Retailers can help to leverage this concept as brand names become associated with reduced risks.

Some factors, such as intellectual property issues, pose challenges to the establishment and maintenance of partnerships. These issues might become increasingly difficult in the future, as such sciences as the microbiology of food continue to advance rapidly.

Despite the challenges posed by the diversity of the players and changing priorities within the system, the potential realm of strong partnerships is large and includes all partners: government, the private sector, consumers, and support players. Partnerships should be formed and function in an open process that is independent and protected from political, economic, and social pressures. Partners must have clear delineations of responsibility and the authority to make decisions to meet responsibilities. They must have the resources to work together effectively. Successful partnerships are based on close, detailed, and accurate communication and collaboration.

In the public sector, the federal government is in the best position to influence how the other components of the food safety system work together. Its actions, which often take the form of regulation, originate in federal food law. The federal government must guide the system with national food law that is clear, rational, and based on scientifically determined risk. A few principles form the basis for ideal food law, which can be conceptualized by the philosophical structure of food safety. This structure includes a process that is fair and open to participation by all without political, economic, or social pressures and that provides for adequate authority and budgetary considerations.

Legislation must be flexible and enforceable, and it must comprehensively address all aspects of the entire system from production to consumption. The authority and responsibility of the federal government and its interface with private and other partners must be well-articulated. Definition of a broad federal role is promoted by the similarities of food safety issues across states and demographic groups. Other levels of government have a role in shaping federal activities if they are to be effective in implementing national standards and in dealing separately with local issues.

With a sound food law in place, partnerships between the federal government and others in the system can establish a framework to provide several important functions. For instance, partnerships of federal, state, and local governments with industry, universities, private organizations, and consumers can ensure that the system is science-based and risk-focused, that surveillance and monitoring efforts provide sufficient information to maintain and improve effectiveness, that research and education efforts are properly focused, that regulation and enforcement are effective and consistent, that the system is responsive to new technologies and changing consumer needs, that a continuous process of evaluation can respond to the queries and problems of all stakeholders, and that resources are adequate and appropriately allocated throughout the system.

Risk Analysis

The scientific foundation for decision-making within a food safety system is risk analysis. The role of risk analysis in an effective food safety system is threefold: it provides a basis for identifying where resources should be allocated in the short term; it constitutes a mechanism for determining where public and private efforts should be directed in the long term, especially with respect to research and preventive measures; and it yields important information for estimating and analyzing the costs and benefits of policy alternatives.

The components of risk analysis-risk assessment, risk management, and risk communication-use interdependent approaches, methods, and models; none can function well in the absence of the others. Good risk communication is required for effective risk assessment and risk management. All components of risk analysis-scientific, economic, legal, behavioural, or other-require an environment that is independent of the derivation of policy and must be subject to agency leadership.

Focused risk assessment identifies risks that have the most important consequences for human health and (not always the same) shows where the most progress can be made with available resources. Scientific risk assessment should identify the risks associated with one or more possible actions, or risks associated with taking no action. Risk assessment is fairly well-developed for chemical hazards, such as pesticides, but not for foodborne pathogens.

Development of risk assessment for the complexity of such pathogens must be given high priority. A pivotal component of risk analysis is risk assessment. Scientific risk assessment is the process of determining the relation between exposure to a hazard at a specific magnitude and the likelihood of an adverse event or disease. The science should characterize the nature and magnitude of risks to human health associated with food related hazards.

On the basis of the scientific assessment, regulators can make informed decisions about where and how to allocate resources to prevent and control hazards. The same scientific technique can sometimes be used for a range of problems throughout the entire system. Uniformity in the process of risk measurement can reduce uncertainty about how specific foods or processes will be addressed, and thus benefit non-regulatory components of the system as well.

Adequate Surveillance and Monitoring

Strong surveillance and technical support provide an infrastructure to set priorities for research, education, and response. Food safety, perhaps more than any other health-related activity of government, demands a continuing and dynamic "moving picture" of the system, rather than periodic "snapshots." For successful leadership, the federal system requires the ability to identify and set priorities among potential food safety hazards and emerging issues. Surveillance and monitoring, education and research, and enforcement and regulation are all used to identify, set priorities among, and address concerns.

Surveillance is the first step toward building the capacity to detect and respond to sporadic instances and outbreaks of foodborne illness. Surveillance and monitoring are also essential for evaluating the system, identifying emerging issues and new trends, and assessing risk from the farm or sea to the table. Monitoring foodborne disease and related hazards can allow early and rapid detection of hazards and illnesses. Thorough investigation of outbreaks to determine their sources and causes can identify crucial weaknesses in the system.

Focused Education and Research

Part of the role of the federal government in ensuring safe food is to promote education and research. Continuous, high-quality research is needed to keep capabilities of regulators up-to-date and to ensure rational and maximally effective evaluation and decision-making as well as optimal preventive measures.

Effective education, the strongest and most important form of information transfer, requires partnerships at the federal, state, regional, and local levels. The federal government should lead efforts to educate and inform the general public on such issues as labeling (including health and nutrition information), standards and specifications, changing technologies, and advertising. All stakeholders in the food system should be fully informed of risks and of food-handling practices that reduce those risks.

Both the responsibilities and the needs for setting education and outreach priorities are shared among many stakeholders. Integration and collaboration can help to focus efforts where the needs are greatest. Priorities for education are closely tied to communication capabilities and should be based on need. General academic and land-grant institutions in partnership with private industry and a variety of organizations play critical roles in the transfer of information. Producers, processors, and consumers have needs for both general and specialized education.

Carefully planned education programmes to communicate fundamental issues are important, with additional opportunities for education in times of crisis. Both should be used during major or highly publicized foodborne illness events. Education about the critical role of consumers in the food safety system might be especially effective.

Research priorities should be focused on the prevention of foodborne illness; this requires research on measures of current problems to determine how the problems can best be brought under control. Specifically, research priorities should be related closely to needs as determined by effective surveillance.

Some research priorities in an effective food safety system must be responsive to emerging problems, but others must have a long-term perspective, and effective research management requires a capability to shift resources and emphasis. In the past, research priorities have often been based on opportunities or pressing issues of the moment. Typically, research funding has been event-driven, rather than evenly sustained in support of longer-term objectives. The press of current problems is an ever-present threat to the needed long-range research, and an effective food safety system will support and protect both. Research must always be relevant to societal concerns and to the complex problems of ensuring a safe food supply. Research priorities should be established internally and externally in partnership with private industry, academe, consumers, and other stakeholders. The priorities must be widely known and accepted; it can be difficult for any agency to maintain an appropriate long-term balance when the news media, the public, legislators, and others, direct issues to be set aside when a crises of the moment arises.

The research budget, especially for long-term projects, should be protected from rapid swings and decreases. Perhaps more than most other activities, the development of effective research teams requires development over time, with significant effort, patience, and resources. A single lean budget can destroy the benefits of effective team-building efforts over a period of many years.

Not every research need can be met, but a force of the best scientific research investigators are needed to solve the long-term, difficult, and highly complex issues of food safety. These experts must be dedicated to the mission of protecting public health by ensuring a safe food supply and they must not be distracted by needs for research in other fields, such as improved methods of food production or processing, or treatment of food related illnesses.

Consistent Regulation and Enforcement

Mechanisms for the prevention of foodborne illness can span the spectrum from voluntary to regulatory and from outright bans to warning labels and education. Regulation is often intended to be a preventive and protective measure. A prerequisite for any regulatory action should be the assurance of adequate, consistent, and effective enforcement. The ability to protect requires the authority and resources to take action. The food safety system must contain adequate provisions for enforcement of regulations and must clearly link responsibilities with accountabilities. Many challenges posed by the current food safety structure, as described in Chapter five of this report, result from an inability to take action, respond to needs, or protect.

Consistency is important in all aspects of ensuring safe food. Similar risks require similar planning, actions, and response. For instance, the intensity, nature, and frequency of inspection should be consistent for foods associated with like risks.

Adaptation to New Technology

The food supply and food production, processing, distribution and consumption practices in the United States have changed dramatically during the past 70 years. Further major changes will certainly occur, though it cannot be predicted when or what they will be. The fundamental shifts from a few, largely local items to today's variety, from small neighbourhood stores to supermarkets, from familiar individual ingredients to prepared foods, from home to outside preparation or consumption, and from traditional to "ethnic" cuisine were not widely predicted before they were well in progress. As these shifts continue to occur, the technologies associated with production, processing, manufacturing, and delivery of food also continue to change.

Hazards associated with food consumption also are continually changing. In recent years, the risks associated with fresh fruits and vegetables have grown because of increased consumption, variety, and importation of produce from developing countries. Importation of different types and amounts of foods from outside the United States expands the federal responsibilities and expertise needed in the US food safety system.

The system must include primary prevention of new food hazards through research and other activities and be flexible to respond to new and potential hazards. The food revolutions of this century have proceeded irregularly and often simultaneously, but few have been adequately anticipated. It is certain that revolution will continue, and an effective food safety system will prepare for the known as well as for future hazards that cannot be clearly seen.

The changing attitudes, concerns, desires, needs, and roles of consumers dictate the characteristics of the food supply and its associated risks. The United States has become home to people of many origins and cultures. With those cultures has come the desire for particular foods. A great challenge to a food safety system is to meet the changing demands while securing safety of a rapidly expanding and diversifying food supply.

Human and Financial Resources

The immense responsibility of an effective food safety system is coupled with the need for adequate human and financial support. If federal and other programmes are to prevent, identify, track, control, and respond to food related illness and reduce risks of future outbreaks or hazards, they will require commensurate funding. Insufficient funding and lack of incentives can diminish an ability to inspect and control foodborne illness; curtail needed research; prevent sustainable, aggressive, and innovative education efforts; and create instability in the system. For instance, the role of states as a public partner in the food safety system has traditionally been important. State and local regulators can help to ensure the safety of food in commerce, and they can assume responsibility for monitoring and enforcing safe food-handling practices; but adequate state and federal funding is needed to carry out these tasks.

The economic costs and benefits of reducing the risk of foodborne illness are difficult to determine accurately and consistently. Regulations that govern the production, processing, distribution, and marketing of food can create benefits by improving food safety, but they can also increase producers' costs and potentially raise food prices. The economic costs of inaction are difficult to assess because the data typically reflect only reported incidents and deaths; unreported events generally elude economic assessment. Net benefits of increasing food safety can be maximized by equating the marginal benefits of safer food with the marginal costs of achieving food safety goals. Market-oriented approaches to food safety include economic incentives. In addition, personnel must be trained, developed, and kept up-to-date with advances and innovations elsewhere. Scientific and managerial competence must be ensured to build the human capacity necessary to support the system.

Provision of adequate funding and staffing are not sufficient; the system must also be efficient and cost-effective. One cost-effective measure is early detection and early warning of outbreaks before they spread. Federal coordination with regard to research and education efforts can ensure rapid and efficient response.

Every system should be practical. The mission must be practical, the goals must be attainable, and the work must be feasible. Primary food hazard prevention approaches and research (both long-term and short-term) efforts must focus on practical applications. Inspection, surveillance and monitoring, and education programmes should be practical in their gathering and conveying of information.

The federal government plays a prominent role in ensuring that vital elements are in place and that the unifying mission of food safety is sufficiently addressed. The coordinated application of scientific, technical, and regulatory measures to prevent foodborne illness and promote public health and safety is critical to the federal government's mission, with its core responsibility to ensure an effective food safety system. There are also occasions where regulatory activity can encourage or require the use of new technologies, which causes movement toward higher standards in the system.

Private Sector Partners

Private partners working in an effective food safety system include producers and importers, processors, marketers (retail and wholesale), food services, trade organizations, professional societies, and private organizations.

The primary and common roles of the various private-sector partners include bringing food to the tables of consumers and bringing sound scientific information to others in the partnership. Within the food production system, the private sector has the primary responsibility for ensuring food safety. To be most effective in this task, the private sector must maintain close interaction with the public sector, which sets standards and provides oversight, and with the consuming public, which expresses its food needs and choices. The roles and contributions of specific private partners are varied and distinctive, and each can make unique contributions to the overall functioning of an effective system. Several examples of successful partnerships that highlight private roles in the framework of an effective food safety system are provided below.

Producers

Production in some areas may represent the least controlled point in the food supply system. However, there are valuable opportunities for effective partnerships between producers and governments in the food safety system. An example of a combination of regulatory and non-regulatory partnership between producers and federal and state governments is the National Poultry Improvement Plan (NPIP), in which state authorities cooperate with the US Department of Agriculture to administer regulations for the improvement of poultry, poultry products, and hatcheries.

In this example, industry participants follow both the general provisions of the plan and specific provisions that are related to their industry category hatcheries, dealers, and breeders. Food animal industries use and understand on-farm quality assurance or control programmes; these could be expanded to include food safety provisions if they are cost-effective and serve as an incentive to market products. The industries are to follow all sanitary provisions and undergo testing and inspection. The industries pay most costs of the plan and seek to achieve and maintain a free or "clean" status. The plan has been useful, especially in breeding flocks, and has helped to eliminate several poultry diseases and prevent the spread of disease from breeders to commercial flocks. More recently, the NPIP adapted an on-farm sanitation programme to prevent and control the serious human pathogen *Salmonella enteritidis* in hatching eggs and chicks.

The NPIP provides a working model that involves federal-state-industry partnership. Industry members use the findings to inform the public that they have met the highest standards and have free, or "clean," status, a designation that adds value to their products and is a "preharvest" animal-production model that might be used in an expanded version for other food animal species. Federal and state agencies benefit from reduced requirements for inspection and oversight of "clean" facilities.

Marketers and Distributors

The dairy processing industry recognized the need for effective control of foodborne pathogens by promoting the enactment of effective controls of milkborne disease, including pasteurization. To assist states and municipalities in their efforts to prevent milkborne disease, the US Public Health Service developed model regulations for voluntary adoption by state and local milk control agencies in 1924. A model milk regulation, the Grade A Pasteurized Milk Ordinance, was developed and implemented through partnerships between the dairy industry and federal, state, and local regulatory agencies.

Private industry can contribute to the shaping of the scientific basis of the future food safety system. For example, the Flavour and Extract Manufacturers Association and the Food and Drug Administration developed the FEMA Generally Recognized as Safe review programme to ensure the safety of food flavours and extracts. In another example of industry-FDA cooperation, the development of low-acid, canned food regulations was a major step toward the safety of canned foods by reducing risks of botulism. Those examples illustrate how a progressive, scientific partnership can contribute to a food safety system.

Education and information transfer can also be important contributions of private industry and its associated trade organizations. For instance, the Food Distributors International and its retail partners have engaged in a cooperative food safety initiative, implemented through the Food Marketing Institute. This project assists the allied food industries to share information, to keep up to date on regulatory developments and technical resources in food safety, and to develop and maintain procedures for the industry.

The egg industry has integrated government efforts, academic research, and other industries to address food safety concerns. The industry has implemented new technologies developed by government and academe to reduce foodborne illness, has developed model partnerships, and has educated and informed consumers about the safety and nutritional composition of egg products.

The Role of the Consumer

Consumers have a large and critical role in an effective food safety system. First, the food system revolves around the purchasing power and decisions of consumers. Annual expenditures for food are enormous—over $700 billion in the United States. More important, the health, well-being, and longevity of the nation's population ultimately depend on individual consumer decisions. It is the role of the private (industrial) sector to deliver to consumers food that is wholesome and safe and food that can be rendered or maintained safe by appropriate handling and preparation by the consumer. After food leaves the production and distribution systems, a different set of risks affects food that is prepared and consumed at home, for which only consumers can be responsible for safety, and food that is prepared by commercial establishments away from home (restaurants, takeout shops, grocery store delicatessens), for which vendors retain major responsibilities for safety, although consumers must still follow good safety practices.

As has been noted before, the food system and the food safety system have constantly changed in many ways to meet consumers' perceived needs. This evolution has depended on close interactions between government and the private sector, but the role of government is often unseen except during crisis situations. Continuing activities of the government and the private sector provide the foundation of a safe food supply, but they must also provide the information and education that consumers need if they are to meet their own responsibilities.

Although consumers' desires, needs, and cultures have changed, the basic standards of food hygiene in the home have changed very little and probably will not change much in the foreseeable future. The food safety system should continue to include fundamental information about frequent and effective washing, prompt refrigeration, and proper cooling. Consumer knowledge on good practices is essential to consumer action, including handling, storage, and preparation of food in ways that improve food safety. Consumers should have reason for confidence that food and food components are safe when received, and they should know how to protect their food as it moves through preparation and consumption.

The size and importance of the consumer interactions with the food system have contributed to the growth of consumer organizations. These play an important role, particularly in advocacy about food safety matters. In addition, they help greatly by providing education and information. They should be involved in the development of protocols and studies to fill the vital information needs of the food safety system. Their leadership role requires them to ensure that their positions on food safety issues are science-based, and that they fairly represent the concerns of consumers in their advocacy role.

The Consumer's Role in Food Safety

In the early 1900s, pasteurization of milk was becoming common. It was the consumer's responsibility to preserve the quality of milk after it was delivered to the home. The consumer was instructed that "as soon as possible after the milk is set on the doorstep, it should be taken in and put in a cold part of the ice box. If some time must elapse between delivery and care by the consumer, an insulated box should be provided to protect the milk from heat and/or freezing. Milk should be placed only in clean pitchers or receptacles. Milk that has stood at room temperature during a meal should not be poured back into cold, fresh milk, but kept in its own container." Keep clean and cold was the rule that was followed to preserve the quality and flavour of milk.

Today, the rule of almost a century ago still applies to consumer responsibility, as is demonstrated by the Fight BAC campaign's educational message. New mechanisms for production, processing, and delivery have evolved, but consumers remain responsible for ensuring that milk and other products are kept clean and cold.

The Role of Other Partners

Government, the private sector, and consumers are not the only recognized organizations that are vitally involved in the food safety system. Scientists, educators, and food specialists are also key partners who are placed throughout the system and in academic institutions. For example, the Institute of Food Technologists represents food scientists, and the American Veterinary Medical Association represents veterinarians in industry, academe, and government agencies. Both professional organizations have subgroups that interact with the government agencies.

The news media cannot be considered an active partner in a food safety system, but are important players. Objective, accurate reporting about food hazards and timely transfer of information to the general public are essential.

The news media also engage in efforts to educate the public about basic food hygiene. In an effective food safety system these groups are recognized for their distinct and important roles and are encouraged to participate by membership on advisory groups, by providing direct input to the regulatory process, or by other actions.

Interdependence

Ensuring safe food is an appropriate role for government, but that role is not exclusive, nor is it a contract between government and the American public to guarantee absolute safety. A strong government presence may be necessary for a strong food safety system but is not necessarily sufficient to create a stronger and better food safety system. An effective food safety system is based on a particular set of attributes and relationships and a focus on improving the public's health and well-being, and the roles of federal, state, and local governments must be well defined, coordinated, and consistent in achieving a single mission.

The public often turns its attention to what government *should* do rather than what government *can* do. But what can government at its various levels actually do to promote food safety, and what conditions must exist to optimize its performance? The answers to those questions are based on defining the functions of the food safety system that can best be performed by each of the actors and on the ability to fill in system gaps in light of the inherent limitations of public service today. What can others or other systems do just as well or better, and should they? An effective food safety system is built on fusing the strengths of multiple players, by linking roles and authorities to responsibilities, and by aligning shareholders' efforts toward a single overarching mission of public health. An effective system emphasizes partnerships new and old, integration, and the need for accountability of all participants.

The food safety system in the twenty-first century should be based on a framework of common goals, measures, rewards, teams, and networks that focus on outcomes. Once the framework is developed and operating to address desired outcomes, and once the attributes of a successful system have been integrated, optimum features of a final organizational design will be more apparent. There is a critical need to connect the food system strategically with an effective food safety component; this cannot be an afterthought but rather must be a planned and purposeful activity.

It is clear that the food safety system must function as a integrated enterprise. That is, it must be agile, fluid, connected, integrated, and transparent. Integration is the antithesis of a system with components in isolation, barriers, and disconnected functions, rules, and policies. One of the most important aspects of an effective food safety system is the ability to assume an active, preventive role, as opposed to a reactive mode, to deal with problems already present. Focusing on prevention of foodborne illness rather than on response to failures is key to identifying and controlling outbreaks of foodborne illness. The system must be built to resolve today's problems but also with a special contemplation and flexibility to address future needs that are not yet foreseeable in detail.

Most important, systems for continuous improvement must be based on information and science. Feedback processes are critical to learning, improvement, and change. The food system requires good data at numerous points to measure results and adapt processes.

A unified food safety mission implies that goals are adopted to protect the public health, reduce risks of foodborne illness, and maintain, build, or rebuild consumer confidence. Means for evaluation and feedback are needed to determine whether the food safety system is meeting its goals. The criteria for assessment of the system outcomes include verifiable measures of progress, such as reduced numbers of foodborne illness outbreaks, decreased incidence of chemical residues in food, and a better understanding of foodborne risks that improves public policy and promotes improved consumer practices.

An Effective Food Safety System

- Should be science-based with a strong emphasis on risk analysis and prevention, thus allowing the greatest priority in terms of resources and activity to be placed on the risks deemed to have the greatest potential impact;
- Is based on a national food law that is clear, rational, and scientifically based on risk;
- Includes comprehensive surveillance and monitoring activities which serve as a basis for risk analysis;
- Has one central voice at the federal level which is responsible for food safety and has the authority and resources to implement science-based policy in all federal activities related to food safety;
- Recognizes the responsibilities and central role played by the nonfederal partners (state, local, industry, consumers) in the food safety system; and
- Receives adequate funding to carry out major functions required.

Recommendations

Congress commissioned the National Academy of Sciences, through the US Department of Agriculture's Agricultural Research Service, to undertake the study that resulted in this report. The charge to the committee was to perform two main tasks: assess the current US food safety system and its effectiveness in addressing the continually changing concerns about food safety; and provide recommendations on scientific and organizational changes needed to ensure an effective food safety system for the present and future generations.

The effectiveness of efforts to ensure the safety of food for US consumers is not solely, perhaps not even primarily, the responsibility of the federal government. An effective food safety system depends on the collective efforts of food producers, processors, transporters, suppliers, preparers, retailers, and handlers; of officials at the local, state, and federal levels; and of consumers who select and prepare food. There are three overriding conclusions that the committee came to as a result of its deliberations:

1. An effective and efficient food safety system must be based on science.
2. To achieve a food safety system based on science, current statutes governing food safety regulation and management must be revised.
3. To implement a science-based system, reorganization of federal food safety efforts is required.

The committee's recommendations focus on two main areas. The first addresses the scientific basis of prudent, cost-effective food safety regulation. The second addresses the legal framework and management structure within which the scientific basis should be developed and deployed. The committee presents these recommendations as an integrated, mutually reinforcing package. Specifically, the committee believes that the development and, critically, the use of a scientifically supportable foundation for regulation is not likely to be achieved within the current scattered organization of federal food safety programmes, where responsibility is dispersed, budgets are separate, a unified mission is lacking, and no single official has formal or public overall responsibility and authority for performance.

More study is needed to determine the best organizational plan for a centrally unified framework for managing the federal system. The committee's recommendations do not assume that the only way to improve food safety is to create a single responsible agency. The committee also recognizes that designing the details of a centrally managed federal system will generate controversy and will in any case take time. Moreover, it should be emphasized that many of the specific elements of its recommendations can and should be implemented without waiting for the creation of a centrally managed federal system.

The committee emphasizes the urgency of reform, both scientific and organizational. Threats to the safety of food consumed by Americans are ubiquitous and, whether or not the threats are growing, they are certainly changing. Diverse observers of the food safety system, including entities attached to both Congress and the White House, have been calling attention to specific gaps and deficiencies for more than a decade. Responsive legislative and administrative actions along the lines of those recommended are long overdue.

Scientific Recommendations

The United States has enjoyed notable successes in improving food safety. An example of these is the joint government-industry development of low-acid canned food regulations, based on contingency microbiology and food engineering principles, that has almost eliminated botulism resulting from improperly processed commercial food. Similarly, the passage of the 1958 Food Additives Amendment to the Food, Drug, and Cosmetics Act of 1938 was a "technology forcing" event that improved the evaluation of the safety of added and natural substances and reduced the risks associated with the use of food additives. With increasing knowledge, many rational, science-based regulatory philosophies have been adopted, some of which rely on quantitative risk assessment. Adoption of this regulatory philosophy has been uneven and difficult to ensure given the fragmentation of food safety activities, and the differing missions of the various agencies responsible for specific components of food safety. This philosophy must be integrated into all aspects of the food safety system, from federal to state and local.

The greatest strides in ensuring food safety from production to consumption can be made through a scientific risk-based system that ensures that surveillance, regulatory, and research resources are allocated to maximize effectiveness. This will require identification of the greatest public health needs through surveillance and risk analysis. The state of knowledge and technology defines what is achievable through the application of current science. Public resources can have the greatest favourable effect on public health if they are allocated in accordance with the combined analysis of risk assessment and technical feasibility. It is important to recognize that limiting allocation of resources to *only* those areas where high priority hazards exist can create another problem: other hazards with somewhat lower priority but with a much greater probability of reduction or elimination will not be addressed due to limited resources. Thus, both the relative risks and benefits must be considered in allocating resources.

6

Values for Nutrition Labeling

GUIDING PRINCIPLES FOR SELECTION

The principal task for the Committee on Use of Dietary Reference Intakes in Nutrition Labeling was to provide guidance in translating the science in the Dietary Reference Intake reports to reference values that could be used for the next revision of nutrition labeling regulations. The previous chapters have provided an overview of the task, history, and present status of nutrition labeling—in essence the context within which the committee conducted its deliberations. This chapter provides ten principles the committee has developed to guide the establishment of updated reference values for nutrition labeling.

The committee's approach to how the DRIs would be used as reference values for nutrition labeling was defined within the sponsors' contract language. In particular, this language specified that the purposes of reference values on food labeling are to enable consumers to compare the nutrient content of different food products and to determine the relative contributions of a food to an overall health-promoting diet. The information in nutrition labeling is not intended to be used to plan individual diets. The committee was to identify general guiding principles for use in setting nutrient reference values for nutrition labeling in consideration of the stated purposes. It was to do this by assessing the objectives, rationale, and recommendations for the methodology to select reference values for the nutritive value of food to appear in the Nutrition Facts box. The committee therefore has developed its recommendations using as its main reference materials the nutrient-specific DRI reports, the DRI derivative reports on applications in dietary assessment and in planning, and the preambles, text, and other background materials of appropriate labeling regulations. The committee presents its recommendations as guiding principles—it does not provide nutrient values. Any numbers in the text related to the guiding principles are illustrative only. It is not the committee's responsibility, or its intent, to make regulatory recommendations. Rather the guiding principles provided in this report were developed as science-based recommendations for the sponsors to accept or reject as appropriate to their own activities.

DEVELOPING REFERENCE VALUES

Using the Percent Daily Value

Section of the Nutrition Labeling and Education Act of 1990 (NLEA) requires that nutrition labeling be designed so that it "enables the public to readily observe and comprehend such information and to understand its relative significance in the context of a total daily diet." The Food and Drug Administration (FDA) developed the percent Daily Value (per cent DV) concept to meet this requirement. The per cent DV was modeled on the "percentage of the U.S. Recommended Daily Allowance," an approach used in the 1973 version of nutrition labeling to help consumers understand and compare the relative amount of protein, vitamins, and minerals in food.

Studies in the United States and Canada do, in fact, support this, although increased educational efforts are needed to optimize its potential use as a consumer tool. The per cent DV was selected after careful study, including consumer research and review of public comments. The committee found the rationale for the use of per cent DV compelling and offers no alternative approaches to the DV concept. The committee recommends that the nutrient content per serving of a food be expressed as a per cent DV whenever it is possible to establish this value for a nutrient. The committee notes that when it refers to the DV throughout this report, it recognizes that the DV is a single term that refers to Reference Daily Intakes and Daily Reference Values, which have distinctly different derivations and scientific bases.

Defining the Population

DRIs have been established for 22 distinct life stage and gender groups. These groups were created because the available data indicated that each group has a unique set of nutrient needs that differentiates it from the others. When using the DRI reports to generate reference values for nutrition labeling of the food supply, the population base needs to better represent the general population through a combination of the distributions represented by these life stage and gender groups. The committee therefore recommends using a base population of individuals 4 years of age and older, excluding pregnant and lactating women, to represent the general population. By the time active children reach 4 years of age, their energy requirements are similar to the energy needs of small, less-active adults. Also, in an earlier review, FDA reported that by 4 years of age children's food-consumption patterns are similar to those of adults. The committee considered whether current scientific information indicates that children in North America are assuming adult eating patterns at a younger age. However it did not find evidence from food-intake studies to support moving this age division for the general population. The committee did identify four distinctive life stage groups that should be considered for nutrition labeling; they are defined in Guiding Principle 8.

Using a Population-Weighted Reference Value

As noted above, a single reference value is most appropriate for the Nutrition Facts box, but this value must be designed to be meaningful for a base population that is 4 years of age and older. Even this smaller base population is comprised of 13 separate life stage groups in the DRI reports, excluding pregnancy and lactation. Although the DRIs can differ for these groups, for many nutrients multiple groups have the same values. Because it is not practical to provide a DV for nutrition labeling for each of the 13 life stage groups, it is necessary to combine the DRIs for the groups to produce a single DV for the general population.

The committee considered a variety of ways to compute the DV and concluded that the most scientifically valid approach was to apply weighting based on census data and the proportions of each life stage and gender group in the overall national population. A DV defined in this way will represent a central value of the requirement for the base population, with individual requirements varying around this value. The details are slightly different for nutrients with an Estimated Average Requirement (EAR), where the distribution of the requirements has been defined; for nutrients with an Adequate Intake (AI), where the distribution of requirements could not be defined; and for nutrients with an Acceptable Macronutrient Distribution Range (AMDR), where the reference values are expressed as a range. The rationale, however, is the same regardless of which DRI is provided: because the groups are represented in the base population in different proportions, the DRIs of the groups should be represented in the DV of the base population in the same proportions.

Developing Reference Values Based on the Estimated Average Requirement

A population-weighted Estimated Average Requirement (EAR) should be the basis for Daily Values (DVs) for those nutrients for which EARs have been identified. The committee recommends that the DVs be based on population-weighted values of the EARs for the different life stage and gender groups. This is because the EAR represents the most accurate reflection of the true contribution of a particular food to total nutrient needs in the general population. A fundamental assumption underlying the committee's recommendation is that the DV (expressed as per cent DV) is intended not only to help individuals compare different products within a food type, but also to help them understand nutrition information about foods "... in the context of a total daily diet". To fulfill this function, the DV must take into account that nutrient requirements differ not only by life stage and gender group, but also within any single life stage and gender group. The best point of comparison for the nutrient contribution of a particular food to an individual's total nutrient needs is the individual's nutrient requirement, which is almost never known, but can be represented by the median of the requirement distribution (EAR). The logic is described in the following paragraphs.

The recommendation that DVs be based on population-weighted EARs arose from the examination of two questions. First, given a distribution of requirements, how should a single numerical characterization be obtained? Second, given a collection of distributions of requirements corresponding to different subpopulations, how should these be combined to produce a single, meaningful DV?

The true requirement of any one individual is almost never known, but it can be estimated from the DRIs. For nutrients for which the distributions of nutrient requirements for particular life stage and gender groups have been characterized, the best estimate of an individual's requirement is the EAR for the life stage and gender group to which he or she belongs. This is because levels of intake above or below the EAR will have a greater likelihood of systematically over- or underestimating an individual's needs. Mathematically, the most appropriate single numerical characterization of a distribution of requirements is typically the median. For symmetrical distributions, the median is equal to the mean.

By definition the EAR is the median of the estimated distribution of requirements for a particular life stage and gender group; therefore the EAR represents the best estimate of the nutrient requirement for individuals within a specific life stage and gender group. The probability that any individual in the group has a nutrient requirement above the

EAR is 0.5. This probability declines as requirement levels rise above the EAR, falling to 0.025 at the Recommended Dietary Allowance (RDA). The RDA overstates the needs for 97.5 percent of the population in terms of a specific criterion of nutrient adequacy. Since the RDA is defined to be 2 standard deviations above the mean, a consequence of the normality assumption is that the RDA is 1.2 times the EAR.

For a nutrient with a normal distribution of requirements and a 10 percent coefficient of variation (CV), the requirements of 95 percent of the population will be within 20 percent (2 standard deviations) of the EAR. Thus the EAR is clearly a better single numerical representation of the requirements for the vast majority of the individuals in the subpopulation than is the RDA.

The second issue in calculating DVs based on the EAR is identifying the best approach for combining subpopulation distributions. Intake levels beyond an individual's requirement have no demonstrable benefit. This argument, applied to the population as a whole, suggests that the DV should be the median of the population distribution of requirements. However in the DRI reports, the requirement distributions are given for subpopulations, not for the total population.

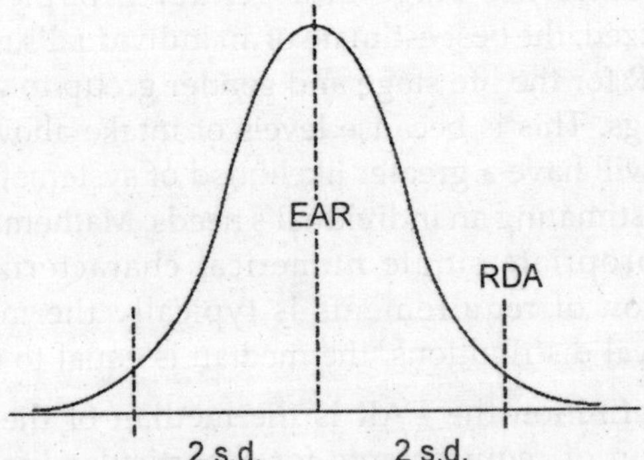

Relationship of the distribution of the population requirements between the Estimated Average Requirement (EAR) and Recommended Dietary Allowance (RDA) for a hypothetical nutrient. Note that 95 percent of the population is within 20 percent of the EAR where 2 standard deviations (s.d.) = 20 percent. The coefficient of variation = 10 percent.

Because it is impractical to provide DVs on the nutrition label for each subpopulation, it is necessary to compute a single number that will summarize the distribution of requirements in the total population. The logic described above argues in favour of choosing a central value of the distribution as the DV. For symmetrical distributions, such as the normal distribution, the mean and the median are identical. However the distribution of requirements for the population, derived from the distributions for the subpopulations, in general will not be symmetrical.

Therefore the median, with 50 percent of the requirements above and 50 percent below, is preferred to the mean, which is sensitive to extreme values of requirements. In summary, the DV should be defined as the median of the population distribution of requirements. This is represented by the population-weighted EAR for nutrients where the distribution of requirements is known. Derivation of this value takes into account the relative proportions of the population in each of the 13 life stage and gender groups that comprise the target population for the Nutrition Facts box and the EAR and the CV of the requirement distributions for each group.

To compute the population distribution of requirements for the DV, the subpopulation distributions are combined using weights obtained from census data. The DV is the median of this resulting distribution. This procedure is easily adapted for different demographic profiles, such as for the Canadian population or for different projected future populations.

Specifically, to calculate the population-weighted EAR for each subpopulation defined by life stage and gender, the requirement for each nutrient is assumed to have a distribution. For nutrients having an EAR, this distribution is assumed to be normal with the median equal to the EAR and a CV of 10 percent. Two exceptions are vitamin A and niacin, which have assumed CVs of 20 percent and 15 percent, respectively. The following text illustrates how the weighting could be approached for nutrients with CVs equal to 10 percent. Slight modifications are required for the two exceptions.

Calculation Examples

As an example, let the population of interest be females and males ages 4 years and older (excluding pregnant and lactating females) in the United States. As stated earlier there are 13 subpopulations with EARs in this population: all children ages 4 to 8 years, and for males and females, separate groups based on the following age breaks: 9 to 13 years, 14 to 18 years, 19 to 30 years, 31 to 50 years, 51to 70 years, and older than 70 years. To calculate the population distribution of requirements, use;

1. The distribution of requirements for each subpopulation,
2. The proportions of each subpopulation in the population.

The DRIs provide the distributions of requirements for the subpopulations. The subpopulation proportions are available from U.S. census data (Population Projections Programme, 2000). The distribution of requirements for the population is called a *mixture* of the distributions for the subpopulations. There are 13 subpopulations; the index *i* with values 1 to 13 is used to distinguish them. Let π_i denote the proportion of the population in subpopulation *i* and let $\Phi_i(x)$ denote the cumulative distribution function (CDF) for the requirements in subpopulation *i*. The quantity $\Phi_i(x)$ gives the proportion of the subpopulation with requirements less than or equal to *x*. The population CDF is thus:

$$\phi(x) = \sum_{i=1}^{13} \pi_i \phi_i(x)$$

The median of the population requirement distribution is the value of *x* where $\Phi(x) = 0.5$. There is no simple formula for this median. However, it is a simple task to calculate $\Phi(x)$

for a very large number of values of x. From these results the value of the median can be determined to any arbitrary number of significant digits. The probability distribution function (PDF) provides an alternative view of a distribution. To denote PDFs, $\Phi(x)$ is used. The relationship between the population PDF and the subpopulation PDFs is similar to that for CDFs:

$$\phi(x) = \sum_{i=1}^{13} \pi_i \phi_i(x)$$

Just as the EAR is the best estimate of an individual's nutrient requirement, there is no single value that would be a better representation of the nutrient requirements of individuals in the population than the population-weighted EAR. Using Indian population predictions for 2005, 54 to 85 percent of the entire population will have requirements that are within 20 percent of the population-weighted EAR, and 72 to 95 percent will have requirements that fall within 30 percent of this value for the list of nutrients examined.

Using Canadian 2006 population predictions, 55 to 86 percent will be within 20 percent and 73 to 96 percent will be within 30 percent of the population-weighted EAR. The observed ranges highlight two important differences among nutrients:

1. The variation in requirements within the life stage and groups, represented by the CV of the requirement distribution, differs among the nutrients,
2. The requirements for some nutrients differ more markedly among life stage and gender groups than do others. For nutrients with considerable variation in requirements within and among gender and life stage groups (e.g., vitamin A), the "spread" around the population-weighted EAR is greater than for those nutrients that have requirements that are less variable (e.g., iodine). Thus it provides a reasonable basis for a DV that individuals can use to evaluate the nutrient contributions of a particular food to the total diet.

The Population-Weighted EAR and the RDA

The committee's recommendation to use population-weighted EARs as the basis for the DVs represents a move beyond past practice in light of new scientific evidence. Past practice based DVs on the highest of the RDAs or Recommended Nutrient Intakes for all individuals in the population. The logic behind this choice was to set a value that was high enough to cover the needs of almost every individual in the population. Because the RDA was set to include a margin of safety, it was considered a prudent choice for nutritional advice for the general public. Furthermore, when the existing DVs were set, the EAR concept had not been developed, and the only quantification of requirements was in the form of RDAs.

In many cases using the highest RDA yields DVs that are so high that they are essentially irrelevant for most of the population. On the other hand, a rationale that has been given in support of using the highest RDA is that there should be some special attention given to the most vulnerable group, which is defined to be the group with the highest requirements, thought by some to be young children. Examination of the DRIs reveals, however, that the group with the highest requirements (with the exception of iron for women of childbearing age) is typically males, including young males. These high intake requirements are based on the rapid growth of this age group. However this group generally has little problem achieving needed nutrient intake.

Another issue is whether a DV based on a population-weighted EAR would facilitate a more meaningful comparison of food vis-a-vis total nutrient needs than would a DV set at the highest EAR, the highest RDA, or a population-weighted RDA. For the purpose of making nutritional comparisons among food products, any reference value would be sufficient, and the concept of a margin of safety or total population coverage is not necessary. However, for the purpose of positioning a food within the context of a total daily diet, basing calculations on a value that includes a margin of safety or covers the entire population would actually distort the overall information. As noted above, within any single life stage or gender group the EAR provides the best estimate of total daily nutrient needs.

The RDA overstates these needs for 97 to 98 percent of the population. Thus a guiding principle for a DV based on the highest RDA for a nutrient would provide an exaggerated impression of total daily needs for most people and would systematically under-represent the true contribution of an individual food to these needs. Using a population-weighted RDA for a nutrient would result in a somewhat lower level than would use of the highest RDA (at least for some nutrients), but it would still be an overestimate of the requirement of most people and an underestimate of the contribution of an individual nutrient to this need. Observations about the implications of the population-weighted approach for nutrient content claims, health claims, food formulation, and overages are included later in this chapter.

It is emphasized that this application of the DRIs is subtly different from the recommended applications for planning diets for individuals. Use of the EAR rather than the RDA is appropriate because the former value provides a better estimate of an individual's true requirement for a nutrient. As such, the EAR provides a better basis against which to appraise the relative significance of a particular food within the context of a total daily diet—which is the goal of the DV. In contrast the RDA is recommended as a goal for planning the diets of individuals. When used as a basis to appraise the nutrient contributions of an individual food to one's total nutrient needs, however, the RDA—by definition—would present an overestimate of needs for most (97.5 percent) of the population. Thus while meeting the RDA may be a prudent goal for an individual's diet plan, the RDA is not the most appropriate measure of need for the population overall.

In summary, an important component of the DRI concept is how each reference value has been derived and its relevance for different applications. For the purposes of nutrition labeling, the committee's task was to provide guidance for the development of a reference number that could be used by an individual to compare the nutrient content of food items within a food type and to place purchase decisions in the context of the food's contribution to his or her total daily diet. The best point of comparison for the nutrient contribution of a particular food to an individual's total nutrient *needs* is the individual's nutrient requirement. It is almost impossible to know the true requirement of any one individual, but a reasonable estimate can be found in the median of the distribution of requirements, or EAR.

The EAR is a daily intake value defined by carefully selected measures of adequacy based on biochemical, functional, or other markers or indicators. As such the EAR represents the best current scientific estimate of a reference value for nutrient intake based on experimental and clinical studies that have defined nutrient deficiency, health promotion, and disease prevention requirements. The EAR, as its name implies, is an estimate of the average of a *distribution of the requirements* for the nutrient in question. For those nutrients for which the distributions of nutrient requirements for particular life stage and gender groups have been characterized, the best, most representative estimate of an individual's requirement is the EAR for the life stage and gender group to which he or she belongs.

Levels of intake above or below the EAR will have a greater likelihood of systematically over- or underestimating an individual's requirement. The RDA is derived from the EAR and is defined to be 2 standard deviations above the EAR on the nutrient requirement distribution curve. Therefore the RDA is not the best estimate of an individual's nutrient requirement. For these reasons the committee recommends the use of a population-weighted EAR, when an EAR has been set for a nutrient, as the basis for the DV. This approach should provide the most accurate reference value for the majority of the population.

Developing Reference Values Not Based on the Estimated Average Requirement

The DRIs are a set of reference values that vary with each nutrient depending upon the scientific information available at the time the DRIs were developed for that particular nutrient. If there was insufficient scientific evidence to develop an EAR for a nutrient for all life stage and gender groups, an AI, an AMDR, both an EAR or AI and an AMDR, or no reference values were developed. For nutrients for which an EAR could not be derived, the committee recommends several different approaches to developing the DVs.

The committee recognizes that the AIs and AMDRs reflect their names in that they do not describe the distribution of intake *requirements* for a nutrient, but rather represent the best approach scientifically available to describe an acceptable intake level or range. Because EARs could not be set for all nutrients, there will have to be a heterogeneity of reference values for the DVs until such time that the science base permits the replacement of AI estimates with EARs. The committee notes however, with the exception of calcium and vitamin D, that EAR values have been set for almost all of the micronutrients that are currently included or are optional in the Nutrition Facts box.

Using Adequate Intakes When There Are No EARs

If no Estimated Average Requirement (EAR) has been set for a nutrient, then a population-weighted Adequate Intake (AI) should be used as the basis for the Daily Value (DV).

Despite the heterogeneous derivation of the AIs, the committee recommends the use of a population-weighted AI for the DV for nutrients for which no EAR exists. Nutrients for which AIs have been set fall into several groups based on the approach used for their derivation: AIs specially derived for infants; AIs based on experimental data (calcium, vitamin D, choline, biotin, fluoride); AIs set using the median intake of the nutrient where no deficiency was observed (pantothenic acid, vitamin K, chromium, manganese, *n*-3 and *n*-6 polyunsaturated fatty acids); and an AI based on the level observed to protect against coronary heart disease (fiber). The AI for fiber is expressed as an amount per 1,000 kcal.

The AIs for infants, which are set for one or both of two life stage groups (i.e., for younger infants ages 0 through 6 months and older infants ages 7 through 12 months), bear brief mention because they were set for specific age categories. For the younger infants the AI is defined as the amount of the nutrient provided in the usual daily intake of human milk; for the older infants the AI is defined as the amount of the nutrient provided by the usual daily intake of human milk and solid food typical for the age group.

The AI was provided for a nutrient if there was not enough scientific evidence available to calculate an EAR. The AI was developed using a "greater degree of judgment than is applied in estimating an EAR" and accordingly there "... is much less certainty about an AI value". These points, along with the heterogeneity of its derivation, make the AI a less desirable replacement for the EAR as a reference value for the DVs. Specifically, the fact that AI estimates do not describe the distribution of requirements for a particular nutrient means that DVs based on population-weighted AIs will not have the same meaning as those based on population-weighted EARs. Insofar as an AI exceeds the mean requirement, a DV based on this value will underestimate the relative contribution of particular foods to total daily nutrient needs. Because the precise relationship between an AI and the true distribution of nutrient requirements is unknown, it is impossible to quantify or adjust for this distortion.

The committee has made its best effort to use the current DRIs for labeling purposes. The lack of an EAR for some nutrients underscores the need for more research in this area to provide the best scientific estimates of nutrient requirements and therefore the best sources of reference values for nutrition labeling. As the study of requirements for nutrients with AIs continues to evolve, it is anticipated that AIs will be replaced with EARs and RDAs. It will be important to then revise the

DVs so that they will all be based on population-weighted EARs and will provide consumers with a consistent standard against which to evaluate the nutrient contributions of a food.

Protein, Total Carbohydrate, and Total Fat

The Acceptable Macronutrient Distribution Ranges (AMDRs) should be the basis for the Daily Values (DVs) for the macronutrients protein, total carbohydrate, and total fat.

An AMDR is not a DRI, but was created to provide guidance for recommended intakes of macronutrients to reduce chronic disease risk. The DRI report on macronutrients established the AMDR and defined it as:

> A range of intakes for a particular energy source that is associated with reduced risk of chronic disease while providing adequate intakes of essential nutrients. The AMDR is expressed as a percentage of total energy intake because its requirement, in a classical sense, is not independent of other energy fuel sources or of the total energy requirement of the individual.

The AMDRs were set because, in the case of some macronutrients or their components, it was not possible to identify a numerical amount where there was a causal relationship between intake and function or criterion of adequacy. Rather, the data better supported a range of intakes that also reflected varying energy needs in the population.

Since there were sufficient data, both an EAR and an AMDR were set for protein and total carbohydrate. Only an AMDR was developed for total fat. The committee recommends using the AMDR to derive the DV for protein, total carbohydrate, and total fat in order to provide a consistent approach that has its basis in risk reduction of chronic disease and healthful dietary practices.

EARs for protein were established for adult males and females based on a rigorous analysis of available nitrogen balance studies. An EAR for protein was established for children ages 1 through 13 years based on a factorial method that adds the amount of protein needed for maintenance based on body weight to the amount needed for protein deposition. The maintenance requirements of adults and the estimates of protein deposition were used to establish the EAR for males and females ages 14 through 18 years. The EARs for protein are expressed in terms of gram per kilogram of body weight and are based on good quality or "complete" protein.

Assumptions about body weight would be needed to convert the EAR for protein into grams per day in order to set a reference value for nutrition labeling based on a population-weighted EAR. Deriving a label reference value for protein based on the new reference weights included in the DRI macronutrient report may not be representative of the requirements of the North American population, which has a high percentage of overweight individuals. Also, a label reference value for protein derived in this manner would likely be below the AMDR of 10 to 35 percent of energy for adults and 10 to 30 percent of energy for older children.

An EAR for total carbohydrate of 100 g/day was set for boys, girls, men, and women of all age groups. The EAR was based on the average minimum amount of glucose utilized by the brain. This level of intake, however, is typically exceeded to meet total energy needs while consuming acceptable levels of fat and protein. Thus using the EAR for total carbohydrate would result in a very low label reference value (e.g., 20 percent of calories for a 2,000-calorie diet), which also would be below the AMDR of 45 to 65 percent of energy for carbohydrate.

An EAR was not set for total fat because there were insufficient data to determine a defined level of fat intake at which no risk of inadequacy or prevention of chronic disease occurs. "AMDRs were estimated for total fat based on evidence indicating a risk for coronary heart disease (CHD) at low intakes of fat and high intakes of carbohydrate, and based on evidence for increased risk for obesity and its complications, including CHD, with high intakes of fat". The AMDRs for fat were estimated for children (25 to 35 percent of energy for ages

4 to 18 years) primarily based on a transition from the high-fat intakes that occur during infancy to the adult AMDR for fat (20 to 35 percent of energy).

To promote healthful dietary practices and nutritionally adequate diets and to provide consistency for setting label reference values for protein, total carbohydrate, and total fat, the committee believes that an approach based on the AMDR is most appropriate. Because the AMDR for each macronutrient is expressed as percent of energy in terms relative to each other, the approach for setting their label reference values should ensure that their sum totals to 100 percent.

The committee recommends using the midpoint of the AMDR for total carbohydrate (since the AMDR for carbohydrate is 45 to 65 percent of energy for all reference groups) and a population-weighted midpoint of the AMDR for total fat (using the midpoint of the range of 20 to 35 percent of energy for adults and 25 to 35 percent of energy for children 4 to 18 years of age). A reference value for protein could then be based on the difference needed for the sum of the macronutrients to equal 100 percent of energy. Using the midpoint of the AMDR as the basis for label reference values avoids extreme values (i.e., lower- or upper-boundary levels) and is an approach that focuses on moderation.

Sugars and Added Sugars

Naturally occurring and added sugars are chemically identical and analytically indistinguishable by current techniques. Naturally occurring sugars (also called intrinsic sugars) are primarily found in fruits, milk, and dairy products that also contain other essential nutrients. Added sugars are defined as sugars and syrups that are added to food during processing and preparation.

The total amount of sugars (in grams) is currently listed in the Nutrition Facts box under the general heading of Total Carbohydrate.

The DRI report on macronutrients established an EAR and an RDA for total carbohydrate; no values were set for either total or added sugars. The discussions of adverse effects of over consumption and hazard identification in the DRI macronutrient report included a complete review of the literature and concluded that the data were not in sufficient agreement to develop a Tolerable Upper Intake Level (UL) for total or added sugars:

Published reports disagree about whether a direct link exists between the trend toward increased intakes of sugars and increased rates of obesity. The lack of association in some studies may be partially due to the pervasive problem of underreporting food intake, which is known to occur with dietary surveys. Underreporting is more prevalent and severe by obese adolescents and adults than by their lean counterparts. In addition, foods high in added sugar are selectively underreported. Thus, it can be difficult to make conclusions about associations between sugars intake and BMI [body mass index] using self-reported data.

Based on the above data, it appears that the effects of increased intakes of total sugars on energy intake are mixed and increased intakes of added sugar are most often associated with increased energy intake. There is no clear and consistent association between increased intake of added sugars and BMI. Therefore, the above data cannot be used to set a UL for either added or total sugars.

The nutrition labeling committee did consider the suggestion in the DRI report about maximal intake of added sugars:

Based on the data available on dental caries, behaviour, cancer, risk of obesity, and risk of hyperlipidemia, there is insufficient evidence to set a UL for total or added sugars. Although a UL is not set for sugars, a maximal intake level of 25 percent or less of energy from added sugars is suggested based on the decreased intake of some micronutrients of American subpopulations exceeding this level.

However it was clear to the committee that the maximal intake level of 25 percent of energy from added sugars, as suggested in the DRI report, would be an inappropriate reference value for nutrition labeling. Such a reference value could be misinterpreted as a desirable intake.

In North America a large and increasing number of adults, adolescents, and children are overweight or obese. The Nutrition Facts box already includes leading information on total calories and total calories from fat. Consumers need guidance about major sources of calories in food, including sugars.

Guidelines for healthy eating, including U.S. government consumer guidelines, often caution consumers to moderate their intake of sugars in general and to sparingly use beverages and food containing added sugars. The major Canadian consumer guidelines are under revision, but a recent fact sheet for educators and communicators that interprets the existing guidelines defines simple sugars and states that "all added sugars, including honey and molasses, contribute primarily energy and taste and have no other significant nutrition advantages". In the United States there is no line item in the Nutrition Facts box for added sugars, and there is no DV for sugars to place this source of energy in the context of the total daily diet.

The nutrition labeling committee considered that consumers attempting to follow dietary advice on added sugars might benefit from nutrition labeling that enables them to easily assess the relative amount and caloric contribution of natural and added sugars in food and supplements. However, without appropriate reference values for total, natural, or added sugars in the macronutrient report, the committee is unable to recommend an approach for developing a reference value for sugars or added sugars for nutrition labeling based on the DRIs. Moreover, it is unclear whether a per cent DV is the most appropriate means for providing information to consumers about sugars or added sugars in the context of a total daily diet. The committee does, however, recognize that consumers need guidance by which to place this important source of calories in labeled food in the context of the total diet. Provision of this guidance should be an urgent consideration of the cognizant regulatory bodies.

The current DVs for protein, total carbohydrate, total fat, and saturated fat are based on a 2,000-calorie reference level. The new Canadian labeling regulations also use this reference level. When the U.S. nutrition label was revised in the early 1990s, a 2,350-calorie reference level was proposed. However the 2,000-calorie reference level was selected because it was thought that a rounded value would be easier for consumers to use and that 2,000 calories was less likely to suggest an inappropriate level of precision. In addition, the use of a lower calorie value was consistent with the public health goals of NLEA. In the United States an estimated 64 percent of adults and 15 percent of children and adolescents are obese or overweight; in Canada it is estimated that 57 percent of men, 35 percent of women, 33 percent of boys, and 27 percent of girls are obese or overweight. Presenting a DV that might further encourage the overconsumption of calories would not benefit the public health of North Americans.

The committee considered whether there was a basis in the recently established Estimated Energy Requirements (EERs) for developing a calorie reference level for macronutrients in nutrition labeling. The committee recognized that using the EER to derive a calorie reference level would require making assumptions about height, weight, and physical activity level. However, the prediction equations used to calculate the EERs were based on normal-weight individuals, but both the American and the Canadian populations have a high prevalence of overweight and obesity. Thus the committee found that the North American data necessary to use the EER concept as the basis for a calorie reference level for nutrition labeling are incomplete and it cannot recommend this approach.

The committee concluded that retaining the current 2,000-calorie reference level would be the best approach as it would provide continuity and would not encourage higher calorie intake and overconsumption of energy. A 2,000-calorie reference level should not be presented in such a manner that consumers construe it to be a mandatory daily intake level for good health. The committee also notes that young children and very sedentary individuals, including the elderly, have energy requirements below 2,000 calories, which underscores the importance of nutrient density in the food consumed by these individuals.

Saturated Fatty Acids, Trans Fatty Acids, and Cholesterol

The Daily Values (DVs) for saturated fatty acids (SFA), trans fatty acids (TFA), and cholesterol should be set at a level that is as low as possible in keeping with an achievable health-promoting diet.

The macronutrient report recommends that saturated fatty acids (SFA), *trans* fatty acids (TFA), and cholesterol intakes should be as low as possible "while consuming a nutritionally adequate diet". In support of this approach the macronutrient report cites research indicating that SFA, TFA, and cholesterol are not required in the diet. The macronutrient report also presents results of regression analyses of various studies that indicate that any incremental increase in intake of these fats correspondingly increases blood total and low-density lipoprotein (LDL) cholesterol and the risk of coronary heart disease.

The committee recommends the application of the DV approach for SFA, TFA, and cholesterol. Use of per cent DVs for these food components would provide a meaningful perspective about their presence in food so that individuals can compare products and make food choices that are consistent with the guidance in the macronutrient report and with the public health goals of NLEA. Inclusion of these food components in the Nutrition Facts box is based on the reduction in risk of chronic disease, and thus for the current nutrition labeling, the reference values for SFA and cholesterol are DRVs.

The committee considered how best to recommend translating the scientific information on SFA, TFA, and cholesterol contained in the DRI report into reference values for the Nutrition Facts box. Since the DRI report did not establish an EAR, an AI, or an AMDR for SFA, TFA, or cholesterol because their presence in the diet meets no known nutritional need, there are no DRI values that can be readily used as the basis for the DVs. Therefore, to establish DVs for these chronic disease-related food components, the committee recommends the use of food composition data, menu modeling, and data from dietary surveys to estimate minimum intakes consistent with nutritionally adequate and health-promoting diets for diverse populations.

Fats are mixtures of fatty acids and all fats contain some SFA. To meet the AMDR for total fat (20–35 percent of energy for adults and 25–35 percent of energy for children ages 4–18 years), some SFA will be present in diets. The question then becomes how much SFA will be present in an achievable health-promoting diet. For example, using menu modeling, diets can be planned that have 3 to 5 percent of calories from SFA. These menu-modeling estimates fall within the recommendations of a report of the National Cholesterol Education Programme (NCEP) Expert Panel on Detection, Evaluation, and Treatment of High Blood Cholesterol in Adults of less than 7 percent of calories that were developed for a Therapeutic Lifestyle Change diet.

Similarly, diets can be planned that provide less than 1 percent of calories from TFA provided that the only sources of TFA are naturally occurring (i.e., in meats, poultry, and dairy products). A recent study that used data from the Continuing Survey of Food Intakes by Individuals reported that the average intake of TFA was 2.6 percent of energy, of which approximately 25 to 26 percent was from naturally occurring sources. A TFA-free diet is not possible if animal-based food is consumed.

For SFA and TFA the committee's challenge was to recommend the best manner in which to use the scientific information in the macronutrient report that would lead to a useful per cent DV. The committee recommends that SFA and TFA amounts be listed on separate lines, but that one per cent DV be included in the Nutrient Facts box for these two nutrients together. The committee recognizes that SFA and TFA are chemically distinct and acknowledges that the macronutrient report identified research that demonstrated physiological effects that differed among the fatty acids.

However both SFA and TFA raise total and LDL cholesterol levels and therefore are potential contributors to CHD risk. Since consumer research has shown that the per cent DV is a helpful tool for comparing different food products that could be optimized further, the committee recommends that the per cent DV be included for both SFA and TFA. By listing SFA and TFA and their gram amounts on separate lines and by providing a combined per cent DV for them, the consumer can be further educated about the unique differences between these fatty acids yet recognize that neither is desirable in terms of CHD risk. As stated earlier one of the main purposes of the Nutrition Facts box is to help consumers compare food products and determine their relative significance and contributions to an overall healthful diet, and providing a per cent DV has been shown to enhance this consumer ability. Further, Health Canada has included this approach in its recent regulations on nutrition labeling.

Providing a per cent DV for combined SFA and TFA on nutrition labeling serves several other purposes. For example, this approach does not promote one type of fat as being more unhealthful than the other. Also, such an approach provides a target and flexible goal for food manufacturers to utilize when combining SFA and TFA in product formulations in order to achieve functional objectives in the sensory appeal and structure of food. Considering SFA and TFA together thus creates an incentive for the food industry to lower both components as much as possible.

With regard to cholesterol the committee noted that a cholesterol-free diet is possible if all animal-based foods are eliminated from the diet; however this is not a realistic dietary pattern for North Americans. An average daily cholesterol intake of 200 mg is attainable if a diet contains two 2-oz servings of lean meats (about 120 mg of cholesterol), 2 to 3 servings of skim milk or fat-free dairy products (about 8–12 mg of cholesterol), and 2 eggs per week (60 mg of cholesterol/ day) as the only major cholesterol sources. Including nonfat-free dairy products (i.e., low-fat, reduced-fat, or whole-fat products), a larger serving of lean meat (e.g., 3 oz), or a third egg per week would contribute additional cholesterol.

The committee recognizes that the dearth of experimental data on acceptable diets that contain minimal levels of these food components makes it difficult to establish DVs for them without further research. The committee recommends that in developing DVs, examples of minimal intake levels of SFA, TFA, and cholesterol estimated through menu modeling should be evaluated against achievable health-promoting diets (identified in dietary survey data) that may be more realistic for a diverse population. While menu modeling provides a basis for evaluating the potential lowest amounts of these fats in a healthy diet, the resulting menus might be well outside the norm for most North Americans. Using dietary survey data will allow these hypothetical menus to be placed in perspective and will allow adjustments to be made that should result in recommendations for meaningful approaches to the intake of SFA, TFA, and cholesterol for the general population.

A DV based on a population-weighted value of the EAR or AI for all life stage and gender groups will reflect the actual contribution of a particular food to the total nutrient needs of the general population. However, individuals in the life stages listed in Guiding Principle 8, have nutrient needs that are physiologically different from those of the general population. A DV based on a population-weighted EAR or AI for the population of people 4 years of age and older would overestimate the nutrient contribution of a food for infants and toddlers and underestimate the contribution for pregnant and lactating women. Therefore the committee recommends separate DVs for food made for these four life stage groups.

Children Less Than 4 Years of Age

In the United States FDA has established substantially different labeling regulations for food manufactured for children under 4 years of age than that manufactured for populations 4 years of age and older. The younger age group is separated into those who are "persons not more than 12 months of age" and those who are 1 to 3 years of age (specifically 13–47 months). In this report, these groups are referred to as "infant" and "toddler," respectively.

Current dietary recommendations are that human milk should be the sole food source for infants until about 6 months of age and should be continued as a milk source until at least 12 months of age. Infants who are not fed human milk, who are weaned before 12 months of age, or who are provided supplemental milk sources before 12 months of age should be fed iron-fortified infant formula. Iron-enriched solid foods are recommended for introduction to the diet for most infants at 6 months of age. In the United States infant formulas are labeled under the implementing regulations of the Infant Formula Act of 1980. Infant formulas are thus covered under separate regulations and do not use nutrition labeling that conforms to what is required for other food.

The final regulation on RDIs and DRVs provides details of the DVs to be used for infants and toddlers. This rule basically uses the highest 1968 RDA for each nutrient listed. Therefore the current infant DVs are the RDAs for infants 7 through 12 months of age. Although indicated as being for infants, the listed RDA actually reflects older infants who receive a mixed diet rather than the exclusively human milk-fed or formula-fed younger infant.

There are several other important differences between nutrition labeling for infants and toddlers and that for the general population. First, protein is listed as a percent of the RDA in nutrition labeling for infants and toddlers, which is not the practice for the general population. Second, saturated fat and cholesterol are not listed in the Nutrition Facts box on food for infants and toddlers. Third, total fat, calories from fat, fiber, total carbohydrate, sodium, and potassium are not given as a per cent DV, but only as a weight of the component. Fourth, the footnotes that appear in nutrition labeling for the general population do not appear on the infant or toddler label.

These differences are designed to ensure that consumers do not improperly focus on the fat content of infant and toddler food and that the diets chosen do not appear to reflect adult caloric density or nutrient distribution requirements. For protein a special rule requires that the protein digestibility-corrected amino acid score for toddlers must be at least 40 percent for the per cent DV to be included in the Nutrition Facts box, otherwise the box must include a statement that the food is "not a significant source of protein". In the United States many foods designed for infants and toddlers list both the infant and the toddler values as per cent DV for protein and micronutrients.

The new Canadian food labeling regulations include different label specifications for children less than 2 years of age. The Nutrition Facts table for food intended solely for children under 2 years of age is specifically not to include: the per cent DV for total fat (or the sum of SFA and TFA), cholesterol, sodium, potassium, carbohydrate, or fiber and the energy values from fat or fatty acids. The Nutrition Facts table must contain the amount of calories and gram amounts per serving of a stated size for total fat, sodium, carbohydrate, fiber, sugars, and protein, with per cent DVs for vitamin A, vitamin C, calcium, and iron.

General guidance for infant feeding has been provided in Canada through a statement of a joint working group comprised of the Canadian Paediatric Society, the Dietitians of Canada, and Health Canada: *Nutrition for Healthy Term Infants*. The statement recommends breast-feeding for at least the first 4 months of life and, for formula-fed infants, cow's milk-based, iron-fortified formulas until 9 to 12 months of age. Labeling, composition, and related packaging and processing of infant formulas are regulated under the Canadian Food and Drug Regulations. Under this law and its amendments, infants are defined as "a person who is under the age of one year," and the nutrient content and composition of infant formulas are tightly regulated. The food label must include:

1. The content of protein, fat, available carbohydrate, ash and, where present, crude fibre,
2. The energy value expressed in calories,
3. The quantity of all the vitamins and mineral nutrients set out in Table.
4. The quantity of choline and of any added nutritive substances contained in the human milk substitute portion of the food, expressed in grams per 100 grams or per 100 millilitres or in International Units for table II nutrients of the human milk substitute portion of the food as offered for sale or in a stated quantity of the food when ready-to-serve food.

The regulations further state that the per cent DV of fat, SFA, TFA, sodium, potassium, carbohydrate, fiber, and cholesterol or the number of calories from fat or SFA and TFA cannot be included on the infant formula label.

Recommendations for Nutrition Labeling for Children Less Than 4 Years of Age

An EAR for toddlers exists for the major nutrients (except for calcium and vitamin D). Because there is a single age and gender group for toddlers, there is no need to use population weighting. Therefore, for nutrients with an EAR for toddlers, the committee recommends that the EAR be used as the basis for the DV; for nutrients where there is no EAR, the committee recommends that the AI be used for the DV.

The situation is more complex for infants as a result of the age split at 7 months, which reflects the change from a virtually exclusively human milk- or formula-based diet to one that includes age-appropriate solid food. An AI was established for most nutrients based on the nutrient intake of infants fed human milk. EARs that were established for some nutrients were specifically for 7- through 12-month-old infants.

At this age, weaning food may provide most of the ingested nutrients (e.g., iron and zinc). For calcium, although only an AI based on the human milk-fed infant is included in the tables of AI values, the DRI text includes different values for formula-fed infants. This is due to the presumed lower bioavailability of calcium in infant formulas relative to human milk.

TABLE: Comparison of Nutrient Reference Values for Toddlers Ages 1 to 3 Years

Nutrient	1968 RDA	1989 RDA	EAR	AI
Iron	15 mg	10 mg	3 mg	N/A
Zinc	N/A	10 mg	2.5 mg	N/A
Calcium	800 mg	800 mg	N/A	500 mg
Magnesium	150 mg	80 mg	65 mg	N/A
Vitamin A	~500 µg RAE	~500 µg RAE	210 µg RAE	N/A
Vitamin D	10 µg	10 µg	N/A	5 µg
Vitamin C	40 mg	40 mg	13 mg	N/A
Protein	15 g	16® g	13 g	N/A

Note: RDA = Recommended Dietary Allowance, EAR = Estimated Average Requirement, AI = Adequate Intake, N/A = not applicable, RAE = retinol activity equivalents.

TABLE: Comparison of Nutrient Reference Values for Infants Ages 7 through 12 Months

Nutrient	1968 RDA	1989 RDA	EAR	AI
Iron	15 mg	10 mg	6.9 mg	N/A
Zinc	N/A	5 mg	2.5 mg	N/A
Calcium	600 mg	600 mg	N/A	270 mg (human milk-fed) 335 mg (formula-fed)
Magnesium	70 mg	60 mg	N/A	75 mg
Vitamin A RAE	~375 µg RAE	375 µg RAE	N/A	500 µg
Vitamin D	10 µg	10 µg	N/A	5 µg
Vitamin C	35 mg	35 mg	N/A	50 mg
Protein	16 g	18 g	9.9 g	N/A

Note: RDA = Recommended Dietary Allowance, EAR = Estimated Average Requirement, AI = Adequate Intake, N/A = not applicable, RAE = retinol activity equivalents.

For infants, as for toddlers, the committee recommends that EARs be used as the basis for DVs for nutrition labeling and that the AIs be used when no EARs have been set. The DRIs include two age groups for infants less than 1 year of age, but no separate values based on gender. Only three nutrients have EARs for infants (protein, iron and zinc), and these were set for the 7- through 12-month life stage group. While there in no need for population weighting of the EAR values for infants because they were set for this single life stage group, the AIs for many of the nutrients differ for the two infant age groups. Values for infants ages 7 through 12 months of age serve as the basis for formula and food labeling. The committee recommends that the infant food label continue to be used and to *only* represent the needs of 7- through 12-month-old infants. During the first 6 months of life, virtually all nutrition is supplied by human milk or infant formulas, and infant formulas are labeled based on the Infant Formula Act and its regulations.

Although AIs are used as the basis for many of the infant DVs, it should be noted that the AIs reflect the intake from the whole diet and are not limited to intake from solid food. The committee therefore encourages continuing educational efforts to ensure that parents understand that human milk or infant formula should be the principal source of most nutrients throughout the first year of life.

It is also important to note that for infants during the first 6 months of life, there are no specific requirements that have been identified for most nutrients beyond that provided by human milk or infant formula. Two exceptions are vitamin D and iron. The American Academy of Pediatrics has recommended that 5 μg of vitamin D (the AI level) be provided to human milk-fed infants beginning in the first two months of life. Beyond that provided by human milk, most infants may not require iron until 6 months of age. However, a substantial number of at-risk infants, such as those born small for their gestational age, may require iron at an earlier age. Therefore although the nutrition label recommendations use the values derived for infants 7 through 12 months of age, they reflect the requirements of younger infants for some nutrients. It is important to note that infants born prematurely or those with special health issues may not have their nutrient needs met by the standard DVs on the infant nutrition label. These issues underscore the importance of the role of the pediatrician, in partnership with the family, in monitoring the early nutritional health and growth of infants. The committee notes that while the historic and current approaches to nutrition labeling for infants and toddlers in the United States and Canada differ, it has developed these recommendations with the anticipation that it will facilitate harmonizing nutrition labeling regulations between the two countries.

Pregnancy and Lactation

During pregnancy and lactation, women have elevated requirements for some nutrients. For example, the requirement for pantothenic acid for pregnant women is 20 percent higher than that for nonpregnant women and for iron it is 172 percent higher. Pregnant and lactating women are in three DRI age groups: 14 through 18 years, 19 through 30 years, and 31 through 50 years. The EARs and AIs for pregnant teenagers (ages 14–18 years) are higher for several nutrients and slightly lower for others compared with older pregnant females (ages 19–50 years). In general, the difference between these age groups are less than 20 percent and range from –17 percent for vitamin K for both pregnancy and lactation, to +16 and +13 percent for pregnancy and lactation, respectively, for magnesium. The only exceptions are phosphorus and calcium, where the EARs for both pregnancy and lactation for phosphorus and the AI for calcium are above 20 percent for teens (82 percent for phosphorus and 30 percent for calcium).

The committee considered creating an additional pregnancy category for teenagers and concluded it was not necessary because of recent statistics that show that birth rates for teenagers in the United States have been on the decline since 1990. For birth statistics, teenagers are divided into three age categories: 10 to 14 years, 15 to 17 years, and 17 to 19 years. In 2002 birth rates for teenagers overall were 28 percent lower than in 1990. The decrease in birth rates reported among the middle-teenage category is more dramatic than the older teens, with a decline of 38 percent compared with 18 percent from 1990 to 2002. In 2002 the youngest age group showed the lowest birth rate in 40 years, with only 7,318 births. Further, the relatively small percentage of teenage pregnancies does not merit a separate DV. If teenage pregnancy trends begin to increase in the future, then the creation of an additional group DV for pregnant or lactating teenagers might need consideration.

Dietary Supplements

The Supplement Facts box should use the same Daily Values (DVs) as the Nutrition Facts box.

The Dietary Supplement Health and Education Act defined a dietary supplement as:

a product other than tobacco intended to supplement the diet that bears or contains one or more of the following dietary ingredients: (A) a vitamin; (B) a mineral; (C) an herb or other botanical; (D) an amino acid; (E) a dietary substance for use by man to supplement the diet by increasing the total dietary intake; or (F) a concentrate, metabolite, constituent, extract, or combination of any ingredient.

The Supplement Facts box must include the nutrients that are required to appear on conventional food labels, any botanicals (including the specific plant part), and proprietary blends by weight. The serving size must be clearly stated on the box. Ingredients for which there are established DVs must be listed first on the box, followed by a horizontal line that separates those nutrient ingredients from ingredients for which there is no DV, such as botanicals. The box must state that DVs have not been established for these latter ingredients, which must be clearly marked with an asterisk.

The committee recognizes that a significant proportion of the population at all socioeconomic levels use dietary supplements, particularly nutrient supplements, as an important part of their total dietary intake. In reviewing the background material and developing its approach to the use of the DRIs for DVs, the committee considered the relevance of the guiding principles for conventional food when considering recommendations for the Supplement Facts box. Since the Supplement Facts box requires the inclusion of the per cent DVs for the nutrients that are mandated for conventional food, the committee recommends that the DVs for dietary supplement labeling should be based on the population-weighted EAR or AI for each nutrient as defined for the Nutrition Facts box. In addition, all other guiding principles for nutrition labeling of conventional food should apply to dietary supplement labeling. For supplement products that are marketed to specific life stage and gender groups, which describes four distinctive life stage groups (infancy, toddlers, pregnancy, and lactation), is appropriate for nutrition labeling of dietary supplements.

Tolerable Upper Intake Levels

The committee discussed various possibilities for ensuring that the UL was considered in nutrition labeling. These discussions included the possible use, in the Nutrition Facts box, of the nutrient's ULs and/or the percentage of the UL that is represented in the product. However the committee agreed that the direct use of the UL in the Nutrition Facts box could be subject to misinterpretation, including the possibility that consumers might view the UL as an optimum or, conversely, a toxic amount. Hence the committee does not recommend including the ULs, their representation, or a statement mentioning them in the Nutrition Facts box for conventional food.

The committee noted that—in addition to being the most scientifically justifiable approach—the population-weighted EAR has the added advantage of providing the widest margin of safety relative to the lowest ULs across all life stage and gender groups. In fact reference values based on the population-weighted EAR would be lower than the ULs for all of the life stage and gender groups used to compute this EAR (i.e., individuals 4 years of age and older, excluding pregnant and lactating women), with the exception of magnesium. The population-weighted EAR may be close to the UL for children ages 1 to 3 years for preformed vitamin A, niacin, and folate from fortified food and supplements and zinc from all sources. The significance of this proximity will need to be evaluated on a case-by-case basis.

Magnesium is one nutrient for which the population-weighted EAR would not be lower than the UL for all life stage and gender groups. However, the UL for magnesium is based on *only nonfood sources that are consumed acutely*, and the criterion used to establish the UL is diarrhea from nonfood sources. Magnesium has never been demonstrated to exert any adverse effect when consumed in food.

Supplements, however, differ from whole food in that they are highly concentrated, bioavailable sources of nutrients. While it is nearly impossible to consume levels of nutrients that approach the UL from nutrients that are naturally occurring in conventional food, there are a few studies that demonstrate ingestion at the UL of certain nutrients from combinations of conventional food and supplements or diets that contain highly fortified food and supplements. In the DRI reports the ULs are predicated on the concern that total nutrient intake should not reach a harmful level. The committee recognizes that dietary supplements provide a substantial portion of total nutrient intake in some segments of the population and contribute to intakes well above the DV and RDA in other segments. The committee is concerned about emerging data, which for the first time are combining nutrient intake from food and supplements and indicate intake levels for some nutrients that closely approach or exceed the UL. To help the consumer place nutrients from supplements into the context of the total daily diet, the committee recommends that the regulatory agencies consider how best to include information about the UL on the supplement label.

Additional Issues

During its consideration of the application of DRIs to nutrient reference values, the committee discussed other issues relevant to these values. These issues are: the inclusion of absolute amounts for micronutrients in the Nutrition Facts box, the use of standardized units of quantity in all aspects of nutrition labeling, and the selection and presentation of required nutrients in nutrition labeling that convey a *positive* public health message and have the greatest public health benefit.

Absolute Amounts for Micronutrients

Absolute amounts should be included in the Nutrition Facts and Supplement Facts boxes for all nutrients.

When FDA issued regulations to implement NLEA, it continued a long-standing policy of not including the absolute amount of micronutrients per serving within the Nutrition Facts box. Food products could, however, include the absolute amounts of micronutrients elsewhere in the food label. The regulations require that micronutrients be declared within the Nutrition Facts box as a per cent DV. FDA chose this approach for several reasons. First, previous research demonstrated that per cent DVs better enabled consumers to understand the relative amount of a micronutrient in a food than did the absolute amount. Second, the Nutrition Facts box was designed to be easy to read, and adding the absolute amounts of micronutrients would make the label more complex. The nutrition labeling did not provide absolute amounts because FDA determined that many consumers did not understand the metric system, and there was no formally voiced dissatisfaction with this approach. The new labeling rules in Canada also state that proposed nutrition labels will not include absolute amounts of micronutrients, although absolute amounts will be allowed for the macronutrients in the core group. The committee considered a number of potential benefits and drawbacks for including absolute amounts in nutrition labeling.

Benefits to Adding Absolute Amounts to the Nutrition Facts Box

Adding absolute amounts for micronutrients to the Nutrition Facts box could provide several benefits. First, public health advice is often given in absolute amounts and not as a per cent DV. For example, advice on calcium intake by health educators and health professionals, national health associations, and government consumer information is given in milligrams. As a result, consumers are not able to easily determine the amount of calcium in a food by reading the Nutrition Facts box since it is listed as per cent DV. By adding absolute amounts, consumers would know the amount of calcium in a food product, yet still be able to use the per cent DV for quick comparison with other products.

Second, including absolute amounts would assist consumers who want nutrient information yet are unable to understand the per cent DVs. In addition, some consumers state that they want both pieces of information—the per cent DVs and the absolute amounts—because they seek different information depending upon the nutrient and the food item.

Third, the food label would be a more useful teaching tool for nutrition and health professionals, especially for teaching persons on special diets. This information would be particularly useful if the food label declared not only the absolute amount of micronutrients and per cent DV per serving, but also the per cent DV for a special group if a food also is being targeted to that group. Nutrition educators contend that the presence of absolute amounts for micronutrients on food labels would make it easier to educate consumers about nutrient needs that are unique to a particular life stage and gender group.

Fourth, absolute amounts and per cent DVs (when they exist) for macronutrients already are required in the Nutrition Facts box. Adding absolute amounts for micronutrients on food labels would make the label more internally consistent in the way information is provided to consumers.

Fifth, absolute amounts and per cent DVs (when they exist) already are required on the Supplement Facts box. Adding absolute amounts for micronutrients in the Nutrition Facts box would make the consumer information for conventional food and dietary supplements consistent.

Finally, one problem in communicating information on food labels is the inconsistency of the terminology used to describe nutrient levels in food. On the front panel, where nutrient information may be provided with a nutrient content or health claim, the level of the particular nutrient is expressed qualitatively or in a relative sense, for example, "good" or "excellent" source or "reduced/less." In the Nutrition Facts box, however, nutrient information for vitamins and minerals is expressed as a per cent DV.

Potential Drawbacks

Adding absolute amounts for micronutrients to the Nutrition Facts box has potential drawbacks. First, adding absolute amounts would require more label space, making the label visually more complex and requiring companies to devote more product package space to the nutrition label or to reduce type size.

Second, the additional information on the label might make it more difficult for consumers to use the label to make healthy food choices. For example, studies conducted by FDA during the design of the Nutrition Facts box found that while consumers preferred to have both per cent DVs and absolute amounts on the label, they did a better job using the label that contained per cent DVs alone. In addition, studies have repeatedly shown that when some consumers see large numbers next to a nutrient, they conclude that there is a large quantity of that nutrient in the food, regardless of the units of measure or the relative amount compared to the DV.

However the overall conclusions that have been drawn based on earlier research typically reflect consumers' use of nutrition labeling without experience, education, training, or guidance. Recent studies have focused on the education of special populations. Training programmes and studies with children and adults with diabetes, patients with chronic heart failure, and clinically obese patients seriously striving for weight loss have demonstrated success in teaching patients to use the Nutrition Facts box to make appropriate food choices. With diabetes education in particular, the focus of training sessions, in priority order, is on:

1. Serving size,
2. Grams of total carbohydrate,
3. Grams of fat. For those diabetic patients who are trained to count carbohydrate grams, there is an added emphasis on grams of dietary fiber in nutrition labeling. For diabetic patients with renal complications, the training also includes a focus on grams of protein, total calories, and milligrams of sodium. In the United States most diabetic training, especially with children, does not use the per cent DV, but rather has the absolute amount as its focus.

The decision to add absolute amounts of micronutrients to the Nutrition Facts box should be based primarily on the information that will enable consumers to make healthy food choices. If making healthy food choices is the primary goal of the Nutrition Facts box, then adding absolute amounts should help achieve that goal. Therefore, the committee recommends that absolute amounts of micronutrients be added to the Nutrition Facts box because this addition has significant potential health value to the consumer.

Units of Quantity

Over time the scientific understanding of micronutrients has grown and the units of measure for expressing micronutrient quantities have changed. The following guidelines were used in deciding what the proposed units should be:

- The unit of quantity for nutrition labeling should be consistent with the EAR or AI. Thus the units for vitamin A, vitamin D, vitamin E, folate, and copper should be changed to reflect the new DRIs.

- Where the current unit is appropriate and consistent with the unit in the DRI report, it should be retained.
- For nutrients where there are no DRI values because the report has not been released (electrolytes), the current units should be retained.

Implications of Changes to the Label Reference Values

In response to the study task and perspectives presented at the workshops, the committee considered several implications of using the population-weighted EAR or AI or making other changes to reference values for food labeling. In particular the committee discussed nutrient content claims, saturated fat and cholesterol claims, health claims, food formulation, and overages. The committee does not intend for this section to reflect an in-depth review of these issues, but rather to highlight several areas where it recommends careful consideration of the impact of potential changes. The resulting numerical values are *illustrative only* because the development of actual numerical values would necessitate discussions and decisions about the selection of the best representative numbers for each variable in the formulas. In addition, decisions about issues such as units, numerical rounding, population estimates, and certain aspects of the calculations would need to be made before calculations could be done to generate the actual numbers.

While outside the direct task of the committee, nutrient content and health claims in the United States are dependent on the DVs. The workshop presentations helped to make it clear to the committee that manufacturers were concerned about the impact of changes in the DVs on the criteria for making nutrient content and health claims.

TABLE: Proposed Units of Quantity for Nutrients

Nutrient Unit	Current	Proposed Unit	Comment
Change needed			
Vitamin A	IU	μg RAE	DRI unit is RAE; carotenes will provide the sole source of vitamin A for vegans (show RAE in footnote)

Vitamin D	IU	μg	DRI unit is μg
Vitamin E	IU	mg	DRI unit is mg; the amount shown should be á-tocopherol
Folate	mg	μg DFE	Because fortified foods contain folic acid, this form should be converted to food folate using DFE calculation (show DFE in footnote)
Choline		mg	AI unit is mg, but UL unit is g
Copper	mg	μg	DRI unit is μg
No change needed			
Vitamin K	mg		
Thiamin	mg		
Riboflavin	mg		
Niacin	mg		Although NE is the DRI unit, the label should only refer to preformed niacin
Vitamin B_6	mg		
Vitamin B_{12}	mg		
Biotin	mg		
Pantothenic acid		mg	Shorten name to pantothenate
Vitamin C	mg		
Calcium	mg		
Magnesium	mg		
Phosphorus	mg		
Fluoride	mg		
Chromium	mg		
Iodine	mg		
Iron	mg		
Manganese	mg		
Molybdenum	mg		
Zinc	mg		
Potential change unknown			
Sodium	mg		Units may change pending release of the DRI report on electrolytes
Potassium	mg		Units may change pending release of the DRI report on electrolytes
Chloride	mg		Units may change pending release of the DRI report on electrolytes

Note: IU = international units, RAE = retinol activity equivalents, DRI = Dietary Reference Intake, DFE = dietary folate equivalents, AI = Adequate Intake, UL = Tolerable Upper Intake Level, NE = niacin equivalents.

New labeling regulations also make the following discussion of the proposed changes more relevant. Nutrient content claims have been permitted for food for special dietary, and for all food meeting the compositional criteria for specified claims. For the first time, amendments to the *Food and Drug Regulations* permit five health claims on food, including a claim for dental caries on the labels of certain chewing gums, candies, and breath-fresheners that contain a specified amount of fermentable carbohydrate.

Nutrient Content Claims

For a food to qualify to serve as a "good" source of a nutrient, it must contain 10 to 19 percent of the DV per reference amount customarily consumed (RACC). An "excellent" or "high" food source must contain at least 20 percent of the DV per RACC, the amount of nutrient per RACC for a food to qualify for a good or excellent/ high claim would be lower in most cases if the DVs were based on the population-weighted EAR or AI than if they were based on the current DVs. The example population-weighted EAR is similar to the current DV for vitamin C and lower for most other nutrients by 22 (folate) to 66 percent (vitamin B_{12}, copper, and iron). Because the units of measure for the DV and population-weighted EAR differ for vitamins A and E, it is not readily apparent how the qualifying amounts for these label claims might potentially differ. Population-weighted AIs for calcium, vitamin K, and fiber may be slightly higher by approximately 10 to 20 percent than the current DVs; the population-weighted AI would most likely be lower than the current DV for vitamin D (~30 percent), pantothenic acid (~52 percent), and biotin (~91 percent).

Currently protein content expressed as a per cent DV and the criteria for protein content claims are based on the amount of protein in a food after protein digestibility-corrected amino acid scores (PCDAAs) are applied. The committee recommends that the reference value for protein be based on the difference between the sum of the reference values for carbohydrate (based on the midpoint of the AMDR for carbohydrate) and fat (based on the midpoint of the population-weighted midpoint of the AMDR for fat for children and adults).

If a protein DV based on an AMDR of greater than 10 percent of energy was adopted, consideration would need to be given to the criteria for expressing protein content as a per cent DV, as well as to the criteria for protein content claims. The committee discussed some of the implications—both with and without PDCAAs—of a 75-g DV on protein label declarations and criteria for protein content claims. Under the current regulations a good source of protein contains at least 10 percent of the DV per RACC. Therefore a good source of protein based on a DV of 75 g would require 7.5 g of protein per RACC.

By way of comparison, a large egg contains 6 g of protein per RACC (50 g), peanut butter contains 8.1 g (2 tbs), and canned navy beans contain 9.7 g (130 g). With or without adjustment for PDCAAs, the egg would not qualify as a good source. Peanut butter would qualify as a good source if not adjusted for PDCAAs, but it would not qualify if adjusted (4.7 g/RACC by the Food and Agriculture Organization/World Health Organization pattern and 5.4 g/ RACC by the Food and Nutrition Board/Institute of Medicine pattern). Canned navy beans would qualify as a good source whether or not PDCAAs were adjusted (7.8 g by both patterns).

In a mixed diet that contains ample protein, the correction factors probably are not important. However the factors would become important when evaluating an individual food's contribution to protein intake—especially in circumstances where the diet lacks variety and is relatively low in energy content (e.g., when meal replacement drinks and bars are used in supplemental feeding or weight-management programmes). Because of the complexities associated with evaluating the contribution of protein to a health-promoting diet, the committee suggests a thorough evaluation of the regulatory and nutritional implications of the use of PDCAAs in this context.

Saturated Fat and Cholesterol Content Claims

In general, the criterion for a "free" content claim is the lower limit of analytical accuracy for a given nutrient, the criterion for a "low" content claim is about 5 percent of the DV, and the criterion for a "reduced" content claim is at least 25 percent less than the reference food. A lower DV for saturated fat and cholesterol may reduce the amounts per RACC required to meet the criteria for free and low claims, perhaps making it more difficult to make these claims about food. It is therefore important to take into consideration that the ability to meet current criteria for reduced cholesterol claims also may be affected by a lower DV for saturated fat.

Health Claims

Each health claim has specific nutrient criteria, among other criteria, for determining the eligibility of a food to make the claim.

TABLE: Illustrative Comparison of the Daily Value (DV) and a Possible DV Calculated Using a Population-Weighted Approach

Nutrient	DV	Good	Excellent
Nutrients that have a Population-Weighted EAR			
Vitamin A	5,000 IU	500	1,000
Vitamin C	60 mg	6	12
Vitamin E	30 IU	3	6
Thiamin	1.5 mg	0.15	0.30
Riboflavin	1.7 mg	0.17	0.34
Niacin	20 mg	2	4
Folate	400 mg	40	80
Vitamin B_{12}	6 mg	0.6	1.2
Copper	2 mg	0.2	0.4
Iodine	150 mg	15	30
Iron	18 mg	1.8	3.6
Magnesium	400 mg	40	80

Molybdenum	75 mg	7.5	15
Phosphorus	1,000 mg	100	200
Selenium	70 mg	7	14
Zinc	15 mg	1.5	3
Nutrients that have a Population-Weighted AI			
Biotin	300 mg	30	60
Calcium	1,000 mg	100	200
Choline	—*f*		
Chromium	120 mg	12	24
Fluoride	—*f*		
Manganese	2 mg	0.2	0.4
Pantothenic acid	10 mg	1	2
Vitamin D[g]	400 IU		
(10 mg)	40		
(1)	80		
(2)			
Vitamin K	80 mg	8	16
Fiber	25 g	2.5	5

Generally a food must be a good or excellent/high source of nutrients associated with risk reduction and a low source of nutrients associated with increased risk. Determination of possible effects on the criteria for sodium-and potassium-related claims is pending the DRI report on water and electrolytes.

General Nutrient Criteria for Health Claims

In addition to meeting specific nutrient requirements to qualify for a health claim, a food must contain 10 percent or more of the DV, without fortification, for one of the following six nutrients: vitamin A, vitamin C, iron, calcium, protein, and fiber.

TABLE: Current Nutrient Requirements for Health Claims

Claim	Nutrient Requirements
Calcium and osteoporosis	High in calcium
Sodium and hypertension	Low sodium
Dietary SFA and cholesterol and CHD risk	Low SFA Low cholesterol
Fiber products, fruits, and vegetables and cancer	Good source of fiber

Fruits, vegetables, grains, and soluble fiber and CHD risk	Low SFA
	Low cholesterol 0.6 g soluble fiber/ RACC
Fruits and vegetables and cancer	Good source (without fortification) of one or more of vitamin A, C, or dietary fiber
Folate and neural tube defects	Good source of folate
Soluble fiber from certain food and CHD risk	Low SFA
	Low cholesterol
	Soluble fiber/RACC on nutrition label
Soy protein and CHD risk	Low SFA
Low cholesterol	
Plant sterol/stanol esters and CHD risk	Low SFA
Low cholesterol	
Potassium and risk of high blood pressure and stroke	Good source of potassium
	Low sodium
	Low SFA
	Low cholesterol

In those cases where the population-weighted EAR or AI is less than the current DV, more food products may qualify for a health claim. A higher DV for fiber, based on the AI for a 2,000-calorie reference value, however, may disqualify some food products from bearing a health claim.

Disqualifying Nutrients

Food that contains more than a specified level of fat, saturated fat, cholesterol, or sodium are disqualified from making a health claim, even though all other criteria might be met. The disqualifying amount is typically 20 percent of the DV. Lowering the DV for saturated fat and cholesterol might make it more difficult for a food to qualify for certain health claims. DVs based on a population-weighted EAR or AI concept or other recommended principles may have mixed implications for claims in nutrition labeling under current regulatory criteria. Regardless, the committee believes that the principles presented in this report provide the most accurate scientific approach to using the DRIs to determine reference values for nutrition labeling.

Effects of Nutrition Labeling on Food Formulation

While discussions about the Nutrition Facts box typically revolve around its impact as a tool to help consumers make more healthful food selections, it must be recognized that the regulations governing the Nutrition Facts box and the associated nutrient content claims also influence the formulation of products. Manufacturers often adjust the quantities of particular ingredients or discretionary fortificants so that their products can be shown in the Nutrition Facts box to have a higher percent DV for some nutrients and a lower percent DV for others, thereby meeting the criteria for particular content claims. Thus any changes to the DV or to the list of nutrients included in the Nutrition Facts box can be expected to have some effect on the nutrient profiles of processed food. Furthermore, implementation of the recommended principles for discretionary fortification is expected to affect the inclusion of nutrients and their amounts suitable for fortification.

Overages

In the United States, for the purpose of determining compliance with nutrition labeling regulations, nutrients added to fortified or fabricated food. A food containing a Class I nutrient is deemed to be misbranded if the amount of the nutrient in a composite sample (collected and analysed in accordance with regulations) is not at least equal to the value declared on the label. This requirement differs from that for Class II nutrients, which are those that naturally occur (i.e., are indigenous) in food. The nutrient content of a composite sample containing a Class II nutrient must be equal to at least 80 percent of the value declared on the label.

In order to ensure compliance with label declarations, fortified nutrients are often added in excess (an overage). The amount of overage to ensure compliance depends on several factors, including the chemical stability of the nutrient itself, the manufacturing process (e.g., where in the process a vitamin or mineral is added; how well the vitamin or mineral is incorporated into the product; the conditions of time, temperature, pressure, and moisture), and the conditions used to simulate abusive handling throughout the distribution and retail chain (because manufacturers cannot control conditions after a product leaves their factories and distribution centres). In the United States reasonable excesses of vitamins and minerals over labeled amounts are acceptable within current good manufacturing practices.

In attempting to comply with the regulation for Class I nutrients, some manufacturing practices may result in unnecessary, excessive overages. Excessive overages would be of concern for those nutrients with a low margin between the DV and the lowest UL and for which a serious adverse effect is the basis for the UL. Even in the absence of the potential for an adverse effect, excessive overages, which may not be captured in food composition databases, complicate the evaluations of nutrient intakes and nutritional status.

Public Health Benefit in Nutrition Labeling

The tone of the message conveyed by the elements in the Nutrition Facts box merits careful consideration because the box serves as an important public health communication tool. When the Nutrition Facts box is revised, the committee suggests that thought be given to the selection, organization, and display of nutrients as these elements may impact the tone of the public health message. The Nutrition Facts box currently can be construed as presenting a negative message because many of the required nutrients that appear in bold print on the top of the Nutrition Facts box are those that consumers are expected to restrict in order to reduce their risk of chronic disease. There is no similar emphasis made by grouping, format, or letter size of those nutrients for which consumers are encouraged to increase their intake (e.g., calcium). The priorities of required nutrient selection, label design, and other factors need to be reviewed in light of the potential positive message tone and educational value that could be presented for nutrients included on the label.

The selection of nutrients and food components to be included on nutrition labeling was primarily based on ameliorating nutritional deficiencies and on illustrating the positive and negative nutrient content of food. FDA critically reviewed these nutrients, modified the list, and placed more emphasis on food components associated with chronic diseases and less emphasis on nutrient-deficiency diseases. In particular the revision placed emphasis on those nutrients that reflected the primary public health objective of a reduction in the risk of cardiovascular disease and the secondary objective of a reduction in the risk of cancer.

Periodic reviews of the key scientific issues of public health significance and whether these issues are being addressed by nutrition labeling will help to maintain the scientific currency of the information provided to consumers. These reviews should include discussions with scientific experts to ascertain if the nutrients listed in the Nutrition Facts box reflect the most current scientific understanding of the nutrition, health, and disease relationships important for public health. Appropriate revisions to nutrition labeling should be considered based on these discussions. While changes in the nutrients required in the Nutrition Facts box can have significant ramifications for food manufacturers, the representation of public health issues and positive health messages only can be accomplished by these periodic reviews and, if necessary, revisions to the list of nutrients required in the Nutrition Facts box.

7

Food and Drug Administration

CRITICAL TO REGULATORY DECISION-MAKING

Americans and not Indians enjoy the safest and most bountiful food supply in the world. This is due in part to advances in science and technology that have enabled the identification of more than five times the number of food-borne pathogens than were known 50 years ago and the development of new solutions to prevent and control these pathogens. Yet each year in our country millions of people become ill and thousands die due to food-borne illness.

Recent trends in food consumption are a major reason that ensuring a safer food supply has become more challenging as Americans both choose a greater variety of foods from a continually expanding global supply and consume an increasing amount of foods prepared outside the home. It is because of the continuing threat of food-borne illness and the increasing complexity of the food supply that the Food and Drug Administration (FDA) and other agencies within the Department of Health and Human Services (DHHS) have deemed the protection of our nation's food supply a high priority. The importance of strong federal leadership in the area of food safety cannot be overstated.

The DHHS is organized to link science to its mission. The FDA is the lead agency for applying the food and environmental laboratory science to support the regulatory and nonregulatory food safety goals. The Centres for Disease

Control and Prevention (CDC) is primarily responsible for the epidemiology and laboratory science to support the infectious and non infectious disease prevention goals, while the National Institutes of Health (NIH) is the premier agency for basic and clinical biomedical research. Together these agencies promote food safety, help prevent food-borne disease, and mitigate the clinical and social impact of infectious and noninfectious illnesses.

Effective and efficient solutions to the numerous public health threats posed by an ever-changing food supply can be achieved mainly through the development and application of sound scientific principles. Building and strengthening FDA's science base is a major priority. The impact of science on food safety policy in the past has been substantial and is illustrated in four main areas:

- Science has enabled the identification of recent technologies to detect new and changing public health hazards in food and FDA has instituted an expedited review process for these new technologies. For example, advances in food virological techniques are improving our ability to detect and combat the presence of food-borne viruses, such as the hepatitis A virus.
- Science has enabled the creation of more effective and efficient approaches to solving public health problems. For example, researchers have devised multiple means for improving the safety of sprouts including the evaluation of a wide range of potential agents as a means of decontaminating both seeds and finished sprouts. FDA developed, in just one year, Good Agricultural Practices for fresh fruits and vegetables. These guidelines are being applied both domestically and overseas. In addition, FDA developed a new approach for enhancing the safety of food imports, placing increased emphasis on evaluating underlying conditions in foreign countries. FDA is also working with the U.S. Customs Service to strengthen protection at the border to block importation of unsafe food.

- Science has enabled the evolution of regulatory approaches to reflect the current state of knowledge. For example, FDA was the first federal agency to issue science-based mandatory Hazard Analysis Critical Control Point (HACCP) regulations, in this case, for seafood. In addition, FDA has proposed to expand HACCP to fruit and vegetable juices and these regulations would include a mandatory performance standard. FDA issued final regulations for warning labels for unpasteurized juices. FDA proposed new regulations to enhance the safety of shell eggs, while also announcing the joint development with the U.S. Department of Agriculture of a farm-to-table strategic plan for egg safety.
- Science has enabled the development of new ways to measure the public health impact of prevention and control efforts. For example, FDA recently completed a consumer study to identify food safety practices that consumers are employing and where trends exist.

These types of impacts are the reason risk assessment is leading the Department's food safety regulations in creating solutions and the reason the DHHS and its Public Health Service (PHS) are organized in a manner that clearly links science to the goal of reducing food-borne illness. The PHS bases its public health policy on the best science available.

Good science is critical to regulatory decision-making and FDA cannot solve this problem on its own. In order to achieve high standards of food safety based in sound science—a public health goal shared by FDA and other government agencies—public and private partnerships must be formed in addition to improving communication and coordination among federal and state agencies. It is only by enhancing the science underpinning decision-making that consumers will receive the level of public health protection that they expect and deserve.

It is widely recognized that science and risk assessment are key to the development of effective food safety policy. However, factors such as statutory authority and the availability of resources, budget and staff often ultimately determine how and when science and risk assessment enter into food safety regulatory decisions.

Over the last century, Congress authorized numerous laws authorizing the work of the three federal food regulatory agencies—the Food Safety and Inspection Service (FSIS), the Food and Drug Administration (FDA), and the Environmental Protection Agency (EPA). These agencies make a wide range of decisions based on the their statutory authority, nature of the food safety problem, time constraints, and resources.

The statutory framework guiding each of these agencies determines the different ways in which science and risk assessments are brought to bear. The FDA and FSIS apply many comparable statutory standards in determining the acceptability of food products, including the adulteration provisions of the Federal Food Drug and Cosmetic Act (FDCA), the Federal Meat Inspection Act, the Poultry Products Inspection Act, and the Egg Products Inspection Act. FSIS is required under the latter three statutes to maintain a continuous inspection programme for meat, poultry, and egg products. FSIS inspectors must determine that these products are not adulterated before moving into commerce. In contrast, no similar requirement exists under FDCA, and therefore FDA has the burden of proving that foods other than meat and poultry products are adulterated. These statutory differences probably related to the mindset of the public and the Congress at the time of the statute's enactment.

The EPA was established long after the establishment of FSIS and FDA. It regulates pesticides under three major federal statutes: the Federal Insecticide, Fungicide and Rodenticide Act; the FDCA; and the Food Quality Protection Act. EPA's main duties under these acts are to register pesticides, prescribe labeling and other requirements to prevent adverse health and environmental effects, and establish tolerance levels for pesticide residues. Due to the requirement that EPA's decisions be based only on public health risk and not economics, the agency's activities are heavily based on risk assessment.

In addition to making food safety policies within the boundaries imposed by these statutes, FSIS, FDA, and EPA are also limited by the amount of information and time available to make these decisions. To illustrate the impact of varying amounts of time and information, food safety decisions can be divided into four major categories. The first category includes the major policy changes or enactment of new legislation that can take years and sometimes decades to accomplish. An example is the focus that FSIS has placed in recent years on reducing the risks associated with microbial pathogens through a comprehensive farm-to-table approach to food safety. A second category is the development of major rules, which can take months or years to complete. Third are the immediate regulatory decisions that agencies must make on a regular basis, ranging from weekly to annual decisions, such as the decisions a manager might make on allocating staff resources and setting priorities. The final category involves decisions that must be made in a matter of hours or perhaps weeks such as enforcement decisions that request the recall of adulterated products or the issuance of warning letters.

The urgency with which these decisions must be made determines the extent to which science and risk assessment methodologies can be applied. The first two categories often allow time for decisions that are based on multi-step, quantitative assessments, involving gathering and evaluating of scientific and economic data relevant to an issue; then presenting the data in an understandable form, crafting different approaches to solve the problem, soliciting public input, and making the regulatory choice. In contrast, rapid decisions are frequently needed in the latter two categories—leaving little or no time to conduct a formal risk assessment. However, in these situations, a qualitative assessment can be appropriate and may be more than sufficient for setting priorities and allocating resources.

In conclusion, there are many types of science-based decisions that food safety agencies must make within specific statutory frameworks and timeframes that result in unique decisions by FSIS, FDA, and EPA. Despite multiple barriers and constraints, food regulatory agencies have been able to accomplish significant improvements under current authority. Examples include the Hazard Analysis Critical Control Point (HACCP) system for meat, poultry, and seafood; a single risk-based pesticide standard for raw and processed food; and the establishment of FoodNet and PulseNet, two hazard identification networks that are playing important roles in outbreak investigations.

Future progress in this area includes the development by the President's Food Safety Council of a full assessment report that will identify the best regulatory solutions for reducing food-borne illness as well as detect both gaps and statutory barriers to the implementation of these approaches. The need for statutory change can then better be determined and, if necessary, legislative principles developed that will assist food safety agencies in creating optimal solutions for resource allocation according to risk.

ROLE OF THE INSTITUTE FOR FOOD SAFETY RESEARCH

The purpose of the Institute is to develop a strategic approach that unifies the research agendas and thus strengthens the coordination of food safety activities in three areas: among various government agencies; between the federal government and the private sector; and between the federal government and academia. The creation of the Institute was driven by the President of the United States and his plan for an expanded food safety research agenda to improve the safety of America's food.

In anticipation of the Institute, an interagency working group on food safety research has developed an extensive plan to achieve its mission of coordinating food safety research activities. Details of the estimated $1 million per year plan include:

- Appointment of up to ten people on a rotating 2-year basis as full time staff for the Institute, headed by a leading food safety scientist, possibly using the mechanism of the Intergovernmental Personnel Act.

- Creation of a new public advisory committee that would incorporate outside expertise into Institute processes.
- Development of a policy and budget committee to ensure connections with pertinent food safety agencies across the Federal Government.

In addition to developing a strategic plan to coordinate research agendas, the interagency group was also tasked with gathering and assessing an inventory of federal food safety research, in particular research funded or conducted by the USDA, the DHHS, or the Environmental Protection Agency (EPA). The assessment was organized in broad terms of the risk paradigm into two categories: the management of food-associated risks and the assessment of food-associated risks. The category on management of food-associated risks was further subdivided into detection and control of food-borne hazards. The category on assessment of food-associated diseases was subdivided into six groups: pathogenicity of food-borne microbes; drug resistance and susceptibility of food-borne microbes; epidemiology of food-associated organisms and diseases; risk assessment methods and data; food handling, distribution, and storage; and finally, economic factors associated with food safety.

Analysis of the research inventory was conducted to identify undue overlap or duplication, future needs and opportunities and, to the extent possible, facilitate agency consensus regarding research priorities. In doing this, the interagency group became aware of three major differences among agencies. First, USDA, DHHS, and EPA use vastly different systems for classifying food safety research. Second, the agencies record information at different resolutions and within different time frames. Finally, agencies have generally incompatible data systems, making interoperability even more challenging.

In conclusion, food safety research is essential for the further innovation of tools that can quickly detect dangerous pathogens and for the development of better interventions and solutions to reduce the risk of contamination of our food. There are increasing opportunities to base policies and practices on science to achieve these goals, and the Joint Institute for Food Safety Research should be a significant asset in that quest.

VEGETABLE SPROUTS

Food and Drug Administration

The microbiological safety of sprouted seeds is a unique problem for the Food and Drug Administration (FDA). It emerged as an issue when the National Advisory Committee on Microbiological Criteria for Food—a multi-agency committee that evaluates scientific issues related to the microbiological safety of foods—identified it as a vector for food-borne disease in a white paper on the microbiological safety of fresh produce.

Sprouts represent a distinct problem as the conditions in which the seeds are germinated foster the growth of bacteria. The presence of pathogens on the seeds or seed surfaces during germination, even at extremely low levels, can be amplified by the sprouting process causing an increased health hazard. To date, no technique exists that can completely sterilize the seeds without destroying them.

Despite the risks associated with sprout consumption, shifts in consumer trends towards more healthful lifestyles over the past decade have resulted in increased consumption of raw sprouts and subsequently a dramatic increase in the incidence of outbreaks due to sprouts. These outbreaks continue in the United States, Europe, and other locations throughout the world.

Historically, sprouts have been implicated in a number of food-borne outbreaks. An elevation in the number of food-borne outbreaks raised FDA concerns about the safety of sprouted seeds. At that time, FDA and the Centres for Disease Control and Prevention (CDC) met with the sprout industry to inform them of government concerns about recent sprout outbreaks and discuss ways to address the problem. FDA issued an interim advisory for high-risk persons about the consumption of sprouts and initiated rapid approval of potential interventions. The same year, FDA also issued a nationwide field assignment to determine current sprout industry practices from seed to final product. Also, they held several public meetings on the safety of sprouts to provide a forum for discussion of the current situation, consumer perspectives, agricultural practices, the state of the science, and possible intervention methods.

One of the immediate concerns that FDA had to deal with in combating the problem was that most of the sprout producers were and remain extremely small businesses, around five employees or less. An inherent problem with this is the fundamental lack of available intellectual or fiscal resources to conduct research or to purchase improved technologies to conquer the problem. In addition, few sprout production facilities are actually registered and thus are hard to locate. To overcome these problems, a formal task force was set up through the National Centre for Food Safety and Technology, the FDA, and the U.S. Department of Agriculture's Agricultural Research Service to coordinate research and develop a mechanism of pooling money from sprout producers to fund the research necessary to solve the problem.

National Advisory Committee on Microbiological Criteria for Foods, at the request of FDA and CDC, completed a thorough safety evaluation and assessment of the current state of science associated with sprouts. Specific recommendations provided by the committee include:

- Expansion of education programmes on the microbiological safety of sprouts for those groups in need of information, ranging from the seed producer to the consumer.
- Increased efforts and more in-depth evaluation on how to limit the potential for contamination at the seed milling stage.
- Multiple treatments of seeds to reduce the levels of pathogenic bacteria prior to sprouting.

- Classification of all sprout production facilities as food processing operations.
- Requirement for testing of sprouts for pathogenic activity with irrigation water prior to harvest.
- Development of technologies appropriate to industry.
- Establishment of an expedited review process to assess the different interventions that have been developed to help seeds or treat sprouts.
- Creation of improved trace-back mechanisms and recalls.
- Development of methods to prevent or retard pathogen growth during germination.
- More comprehensive evaluation of seed characteristics.

This review of the current science associated with sprouts illustrates the general framework that FDA uses to deal with many emerging health problems. First, the agency focused on the primary goal of safeguarding public health. Second, the agency gathered the best science and scientists available to evaluate the problem and identify solutions. Third, FDA responded to the problem by creating solutions appropriate for the level of risk and resources involved. Finally, FDA actively involved stakeholders at all levels, and encouraged innovation.

Currently, FDA continues to resolve the distinct food safety and resource issues of the sprout industry. As surveillance and evaluation of sprout facilities continues, the agency will keep generating alerts to the industry and consumers about emerging risks associated with sprouts, provide guidance on ways to reduce those risks, and maintain involvement with stakeholders on a routine basis in order to foster new ideas and solutions. The FDA recognizes the complexity of dealing with food safety hazards and will continue to provide the public with the safest food supply possible through application of science-based solutions to emerging food safety problems.

MEAT AND POULTRY HAZARD ANALYSIS

The Hazard Analysis Critical Control Point approach is a system of preventative process controls that is widely recognized by scientific authorities nationally and internationally and used throughout the food industry to produce products in compliance with health and safety requirements. About five years ago, the U.S. Department of Agriculture decided to apply HACCP to meat and poultry products as a result of a major policy change that focused USDA efforts more effectively on risks associated with pathogens. The implementation of the system is expected to be complete in January 2000. Since the establishment of this rule, the incidence of food-borne illness associated with meat and poultry products continues to be monitored and impacts thus far appear favourable.

The need to focus more heavily on pathogenic microorganisms, and to implement preventive approaches such as HACCP, was established and supported by studies conducted over the past 15 years by the National Academy of Sciences, the Government Accounting Office, and the USDA. The Council for Agricultural Science and Technology estimated that 6.5 to 33 million cases of food-borne illness and up to 9,000 deaths occur each year because of food-borne illness and related problems. However, public support for change in the food safety system did not truly begin to emerge until outbreak of food-borne illness associated with *Escherichia coli* O157:H7 in undercooked hamburgers. Thus, a comprehensive strategy for change was developed with HACCP and pathogen reduction as the centrepiece.

The pathogen reduction and HACCP rule consists of four mandatory provisions. First, it requires all plants to have standard operating procedures for sanitation. The second provision requires slaughter plants to test carcasses for generic *Escherichia coli,* an indicator of fecal contamination. Third, all meat and poultry plants must implement HACCP systems as a means of preventing or controlling contamination from pathogens, as well as other hazards. Under HACCP, plants identify and evaluate the hazards that could affect the safety of their products and institute controls necessary to prevent those hazards from occurring or at a minimum, keep them within the acceptable limits. Finally, to make sure HACCP systems are working as intended, the rule mandates performance standards for salmonella at slaughter and grinding plants.

One of the key guiding principles of the rule was the clarification that industry is responsible for producing and marketing products that are safe, unadulterated, and properly packaged and labeled. With industry assuming its proper responsibility, federal agencies can use limited resources more efficiently and effectively. Another key concept was the combination of HACCP and performance standards for pathogen reduction. HACCP has to be combined with objective means of verifying food safety compliance. Up until this time, microbial performance standards for raw products, with the exception of *Escherichia coli* O157:H7 in ground beef, had not been established. The development of these rules was broadly supported by the public through numerous public meetings soliciting input.

Although the implementation of HACCP is still underway, preliminary data indicates positive results. Salmonella prevalence in broilers, swine, ground beef, and ground turkey was significantly lower after HACCP implementation than in the baseline studies that were conducted before implementation. Data released this year from the FoodNet Active Surveillance System for food-borne illness, the rate of campylobacter, salmonella, and cryptosporidium infections declined nationwide, and that *Salmonella enteritidis* infections declined in all states but Oregon.

Concurrent with HACCP and its implementation within meat and poultry slaughter and processing establishments, USDA's broad food safety strategy addresses every step in the food production process, from animals on the farm, to slaughter and processing, and to product distribution and preparation. The USDA is working through the Partnership for Food Safety and Education and the Fight BAC Campaign to ensure that consumers know how to properly prepare, handle, and store foods. Additionally, USDA is implementing many regulatory reform initiatives to convert traditional regulations to performance standards, thus clarifying the roles and responsibilities of industry and allowing them more flexibility to develop and introduce new technologies to improve food safety.

Many lessons have been learned throughout this process. First is the need to base changes on the best science available and then make adjustments as new information becomes available. Second is the importance of making HACCP mandatory for all plants to ensure that consumers receive safe food regardless of the size of the establishment from which it originates. The third lesson learned is the need to implement HACCP in the context of a broad farm-to-table strategy to enable the far-reaching changes that are necessary to effectively reduce the incidence of food-borne illness. Finally, it is important to have public participation in the process to ensure the discovery and application of the best possible scientific solutions available.

HACCP, as part of a broad-based food safety strategy, is working as intended to reduce the incidence of food-borne illness associated with meat and poultry products. Although many scientific gaps continue to exist that prevent the establishment of "pure" public health standards, revisions in the standards will occur, as new data becomes available. For the future, USDA plans to improve efforts to quantitate the public health risks associated with certain pathogens and foods and ensure that ongoing research will provide sufficient and accurate information on which to base regulatory decisions and develop new preventative procedures. To the extent possible, regulators must continue to focus efforts on designing policies and focusing resources on the most immediate and significant public health risks.

Escherichia coli O157:H7 and Listeria Monocytogenes

Over the past two decades, the Centres for Disease Control and Prevention (CDC), the Food and Drug Administration

(FDA), and the U.S. Department of Agriculture (USDA) have been conducting investigative studies to determine the origin and methods of prevention for two food-borne pathogens, *Escherichia coli* O157:H7 and *Listeria monocytogenes*. Many technologies have been developed, preventive processes implemented, and lessons learned throughout this process.

Following are case studies on these pathogens, with brief overviews of what the agencies discovered, how they responded to these findings, and ideas about what should be done in the future to prevent and reduce the incidence of these and other food-borne illnesses.

Escherichia coli O157:H7

In a series of similar outbreaks over the next 10 years, CDC was able to determine several characteristics about the organism, most importantly that cattle were a principal source or host carrier and that contaminated cattle manure was likely the source of many infections. Studies also revealed that *Escherichia coli* O157:H7 caused hemorrhagic colitis, had no unusual heat tolerance and could be controlled by proper cooking temperatures, had a very unusual acid tolerance, and could survive fermentation of meat.

After a serious outbreak involving more than 700 illnesses and four deaths from eating undercooked hamburgers, that FDA changed the Food Code, which previously recommended cooking temperatures for ground beef that were insufficient to kill large populations of *Escherichia coli* O157:H7. The new Food Code criteria indicated that ground beef patties be cooked to an internal temperature of 155 degrees for 15 seconds. In addition, USDA required that safe handling labels be used for raw meat and poultry products. A year later, USDA declared *Escherichia coli* O157:H7 an adulterant in raw ground beef, established a zero tolerance, and initiated end product testing for raw ground beef. However, food-borne illness continued to increase and it was clear that further action was necessary.

FDA approved irradiation of red meat, and CDC introduced the FoodNet system, an active surveillance system that monitors the occurrence of illnesses associated with selected food-borne pathogens. Additionally, CDC, USDA, and the food industry initiated the Fight BAC campaign, a programme which educates consumers about proper food handling techniques in the home. USDA published a document recommending that a thermometer be used to measure the temperature of cooked ground beef patties. Currently, regulatory agency efforts to reduce food-borne illness due to *Escherichia coli* O157:H7 continue, with USDA planning to implement HACCP in all meat processing plants by the year 2000.

Listeria Monocytogenes

CDC published a case-control study on sporadic cases of listeriosis which revealed that about 20 percent of the cases were attributed to consumption of hot dogs or undercooked chicken. In addition, inoculation studies revealed that *Listeria monocytogenes* can multiply rapidly in certain ready-to-eat processed meat products at refrigeration temperatures and that growth is prevented or delayed in highly acidic foods such as summer sausage. As a result, USDA initiated an end-product testing programme for ready-to-eat meats. This programme was shown to be ineffective as a case of listeriosis expanded its finished product testing to more products and increased the sample size of the foods from 1 gram to 25 grams.

In 1998, a major outbreak of listeriosis occurred involving 101 cases associated with frankfı rters and possibly deli meat. As a consequence, USDA advised meat processors to reassess their HACCP plans and critical control points to include measurement of *Listeria monocytogenes* levels on source materials, further classification of growth characteristics of the pathogen, and re-evaluation of processes thought to kill *Listeria*. USDA also provided guidance to meat processors, recommending both environmental contamination and end-product testing as well as increased educational efforts targeted at high-risk consumers.

Lessons Learned

Listeria monocytogenes and *Escherichia coli* O157:H7 are dangerous food-borne organisms requiring the institution of many preventive efforts. Many lessons have been learned and will continue to be discovered from trying to control such pathogens. Many opportunities for improvement emerged upon evaluation of the regulatory actions taken to reduce the incidence of food-borne illness in these two cases. These include: addressing the limitations of end-product testing, defining critical control points that have the greatest impact on product safety, applying the majority of inspection resources at these critical control points, increasing safety efforts at processing facilities where critical control points are not well-defined, and identifying more effective ways to educate consumers.

Although many of the preventive efforts that were implemented previously are beginning to appear effective, there are still many opportunities to improve the current state of food safety. These include: introducing HACCP from the farm to the table, developing and implementing effective and practical critical control points for on-farm use by producers, and overcoming the major limitations with end product testing. End product testing is a particular problem because if entire lots of product are not retained by food producers, a major portion of the contaminated lot will have been consumed by the time the product is recalled. Furthermore, because of the sporadic distribution and low level prevalence and concentration of pathogens in these foods, end-product testing may frequently not detect contaminated product. However, advancements in technology and science will facilitate further improvements in these areas.

Items to consider for the future include: defining the level of danger at which corrective federal action should be initiated, identifying and developing the roles and responsibilities of industry in solving the pathogen problem, and finding out what treatments or practices are useful for food processors and food service establishments to substantially reduce the risk of pathogen contamination of foods. Most importantly, agencies need to determine the level of responsibility consumers are willing to take in adopting safe food handling practices. It is only through the implementation of preventive approaches from farm-to-table that food safety hazards will be eliminated.

1996 Food Quality Protection Act

A brief review of the statutory history indicates that the first time health was mentioned in relationship to food safety occurred in the 1954 Miller amendment to the Federal Food Drug and Cosmetic Act (FDCA). This amendment required manufacturers to submit health and safety data prior to the registration of a pesticide used on raw agricultural commodities. In 1958, the Food Additives Act was passed to regulate pesticides in processed foods. In 1993, the National Academy of Sciences (NAS) published its report, *Pesticides in the Diets of Infants and Children,* a significant event leading up to the Food Quality Protection Act. In this report it was pointed out that children differ from adults quantitatively and qualitatively in their toxicologic responses to synthetic chemicals and exposure to pesticides. It was also suggested that estimates of exposure to chemical pesticides should include both dietary and nondietary exposures and that tolerance-setting for agricultural chemicals should consider both health and agricultural factors.

Many questions have been raised because the FQPA is fairly new and implementation is just beginning. The FQPA will stimulate the accumulation of large amounts of data for the reregistration of pesticides, the registration of new pesticides, and the reassessment of tolerances. Stiffer data requirements for registration and the cancellation of existing registrations will stimulate the search for more environmentally friendly, less risky pesticides. Revisions to the application rates of pesticides, for example, are another approach that could markedly reduce exposure to and levels of residue. Developments in agricultural biotechnology will allow plant breeders to develop pest resistant varieties that express their pest resistant properties for only a short time or under limited circumstances. The encouragement of biocontrol measures and agricultural biotechnology holds immense promise to further reduce dependence on synthetic chemical pesticides. The FQPA is likely to accelerate research and development.

The opportunities for science to inform the legislative and policy environment and the whole spectrum of food production from farm-to-table are enormous. Science can contribute importantly to the implementation of policy, not just FQPA, but other food safety policies as well. Science can help develop better models for estimating exposures to and risks from chemical pesticides, develop alternative use patterns for existing and emerging chemical pesticides, and provide data on the environmental fate of pesticides and hazards associated with breakdown products of agricultural chemicals. Science is already providing impetus for the development of safer pesticides, with lower risks to health and the environment. Finally, science is providing umbrella opportunities for greater harmonization of food safety policy among regulatory agencies.

8

Agriculture and Rising Food Insecurity

DIMENSIONS OF GLOBALIZATION

According to UN-DAW, "Globalization has become the catch-all term used to refer to those various phenomena and processes that are brought about by changes towards world economic integration. It therefore lacks a neat definition". Its economic dimensions, however, cover the closely related but distinct concepts of openness or liberalization, integration and interdependence of nations.

From this statement, many faces of globalization are widely recognized with varied expectations. These include capital expansion, trade expansion or trade liberalization, cultural integration, financial liberalization, increased information and technology flows, increased labour mobility, changing consumption patterns and so on. Central to all these is increased exchanges or trade. Thus trade liberalization is one of the most touted features of globalization.

In the agricultural sector or in other primary production or craft economies, trade liberalization was jumpstarted by structural adjustment policies (SAPs). To these economies, SAPs continue to be the major face of globalization and has been analysed more than other facets. In this regard, it is always difficult to disentangle trade effects from other globalization effects. In these discussions of the micro level analysis of globalization, we shall also focus on the liberalization policies that have shaped agricultural growth process in Africa.

Central to this is the theory of inequality as it affects unequal partners. Globalization effects are expected to be positive for all trade participants, and although there will be net gains and net losses; it is propounded that the net gain will outweigh the net loss. But in recognizing the inherent equity considerations in the response to the same set of opportunities by different (unequal) actors, it remains doubtful that the net gain will augur well for long term human development especially in poorer countries or between different groups within the same country. According to Elson, adjustment means change and change means costs as well as benefits, losers as well as winners. Change must therefore be managed so as not to leave inequities in is distribution of costs and benefits. And according to ADB, all must be carried along into the income-growth process that liberalization offers, if Africa is to achieve the self-sustaining growth that has always eluded it. As stated by Elson and quoted in Gladwin-

"And if greater reliance is to be placed on private enterprise, we need to ask, whose enterprise? The enterprise of the woman farmer (*on whom household subsistence is hinged)* or the enterprise of (*male-managed, household, market-bound*) agribusiness and merchant with monopoly power? The enterprise of women cooperatives or that of a multinational corporation?"

These diametric concerns symbolize the lack of synergies between the macro and micro impacts of liberalization and central to this is the creation rather than the eradication of (feminized) poverty and food insecurity in Africa.

POVERTY AND FOOD INSECURITY IN AFRICA

In sub-Saharan Africa, a large and increasing proportion of the population subsist on per capita income of less than one dollar a day. The share of the population falling below the poverty line is as high as 50 per cent. Although all indices of poverty are well manifested in sub-Saharan Africa's Human Development Indices, the heart of the problem is food insecurity. From statistical projections, aggregate cereals demand and supply balances for African countries indicate a likely increase in imports from 9 million metric tons in 1990 to 27 million metric tonnes by 2020. And given the obvious difficulties in mobilizing resources to finance imports and the implications on local food availability, deterioration in the food security will result, unless revolutionary departures are made from current production patterns.

Agriculture is not only the primary source of food in Africa; it is the principal means of livelihood in its predominant rural settlements. The challenges then are, not only to drastically reduce net import demands and make agriculture a major source of export earning, but more critically, to achieve agriculture-led industrialization towards the attainment of structural reforms as are called for by increasing world integration. Although the problems are both policy-induced and structural in nature, our concern is on the structural challenges that tend to constrain the desired social and economic transformation of agriculture and that reinforce poverty and food insecurity.

A look at Measurement Issues

The evidences of poverty and food insecurity have never been so clear and alarming in their proliferation as in the last decade. But more worrisome is the conflict in the manifestations. The World Food summit secured international commitment, in 1990, to reduce the number of undernourished people by half by the year 2015. Five years on, statistics show that an increasing number of people remain food insecure (800 million by 2000). Yet this contracts with the supposed progress being made to reduce by half the world's population living in absolute poverty over the same period, which indicates that the international development community is on track. This implies that different indices are being measured or equally likely that progress in one area is negatively affecting the other. So, is hunger or food insecurity a symptom of poverty and in what ways are they linked?

The above paradox makes a clear distinction between food production and food access. Here is where poverty comes into the equation of food security. A poor nation may increase its food production, national food self-sufficiency and economic growth to lift itself upward in the poverty statistics, but sections of its people may remain food insecure because of other factors that affect their access to the food. Suffice it, then, to say that development indices that focus on increased production of food alone are inadequate to capture the pattern of its distribution between populations and within populations.

Poverty characteristics are important in the construction of poverty indicators. Therefore, the relevance of human development indicators (HDI) rather than money-metric or income measures, GDP measures or other macroeconomic structure have been seen to be more appropriate (Human Development reports). The characteristics of the poor are also contextual and so are the manifestations of poverty in different populations. The need for the voice of the people in poverty assessment and the design of poverty reduction measures therefore remain very valid. The PRSP strategy is aimed at streamlining the objectives of macroeconomic growth with those of human development. It also emphasizes the contextual sing poverty knowledge and reduction strategies.

The gender dimension of poverty is particularly reinforced in agrarian economies where the poor are characterized by landlessness, invariable (inelastic) supply of labour, likely to live in female-headed households, likely to be farm labourers rather than farm owners, remote form development assistance due to time constraints, hunger as well as literacy and health constraints and so on. As a result, the consonance of gender development indicators with other human development measures is a necessary condition in poverty assessment. The relationship of growth with poverty has been shown not to be a foregone conclusion, as nations, which recorded high economic growth especially during structural adjustment, also recorded high levels of poverty incidence and high-income concentration. The distributional aspects of growth are therefore more relevant to the measurement of poverty. This view is reinforced by the fact that there is considerable correlation between the trends in Human Development Indicators, Gender Development Indicators and Human Poverty Indices.

TABLE: Poverty Indicators for Selected Countries

	Human Development Indicator	Gender Development Indicator	Human Poverty Index	Macroeconomic Growth (GDP per capita)
$/year				
Norway	0.934	0.932	3.9	36806
Thailand	0.745	0.766	18.7	2593
Ghana	0.556	0.552	35.4	399
Nigeria	0.439	0.421	37.6	256

The Table above shows dimensions of the linkages of gender development with growth and poverty reduction. It shows synergy, not only between growth and poverty reduction, but also between the human development indices and gender development indices. This implies that growth is a necessary but insufficient condition for poverty reduction. Sub-Saharan Africa has lagged behind both in its macroeconomic structure, human development and gender development indicators, compared with other developing and developed areas. It would then be tempting to say that the problem of poverty in SSA is induced both by slow growth and social inequalities and that pro-growth policy like liberalization are insufficient to achieve balanced and sustainable growth with equity. Liberalization policies, in their standard prescription will need to be looked at through a different kind of lens.

Expectations from Liberalization for African Agriculture

The perspective goal of market liberalization worldwide and especially in Africa, as embodied by different forms of economic (structural) adjustment, is *structural transformation (ST)*. This is hinged on market-led growth strategies rather than protectionist 'soft' government for Africa's predominant agricultural sector.

ST entails the development of manufacturing and service sectors, such that the relative importance of agriculture declines (over time). That is, the percent of labour-force in agriculture and the percent of GNP from the sector decline as labour specialization proceeds. An interrelated set of changes in economic structure, including internal consumption, production mechanism, external trade and capital flows, domestic savings and investment behaviour are necessary mechanisms that must catapult the primary production system into a higher technological realm which is required to sustain industrial development. In short, globalization as epitomized by liberalization policies is expected to be the cure-all for Africa's poverty and food insecurity, if and only if Africa's economic structure can be modified towards more efficient production system, systemic industrialization, modernization of agriculture via increased commercialization, specialization, based on comparative advantage and a gradual shift from agricultural-led to industrial-led macroeconomic development.

In assessing the linkages, Kanji identifies two main approaches, which may yield different results. Mainstream economic approach and Socio-economic approach. The former takes the neoclassical view of liberalization whereby potential effects of SAPs for instance are beneficial:

- Free movement of goods and services
- Increased specialization based on comparative advantage
- Increased allocative efficiency in resource use
- Increased technical efficiency by using more capital and technology rather than inefficient labour
- Enhanced production towards export-led growth
- Appropriate pricing leading to positive supply response

- Increased aggregate production leading to increased food security
- Increased farm income leading to poverty reduction (at least in money-metric terms).

However, these standard response parameters do not consider the way that policies are specified and their differential outcomes, if all cannot respond according to these standards. As a result, the poor and other vulnerable groups may not benefit. Also the perspective of food security is over-generalized. Food security is not about the food or the commodity but about people and hunger. Nominal production of food does not guarantee affordability to the poor or accessibility to non-producers. In the last two decades, world output of food has doubled but the world's population of the hungry has also doubled. As empirical proof, SSA's dependence on trade, as measured by share of GDP from trade increased from 38 per cent to 43 per cent between 1988 and 2000.

At the same time, her share of world trade had declined in real terms. Its dependence on primary goods still remained above 80 per cent from oil and non-oil, mainly agriculture. Price volatility of primary commodities led to declining terms of trade which was 21 per cent below its 1970 level. Over the same period, the percentage of poor people remained the same in 1998 as it was in 1987, in spite of increased trade volume. Social indicators also showed a decline, going by HDI estimates. The conclusion is that growth effects need to consider other related social variables, mainly distribution effect of increased income. Growth in the poor's income need to be decomposed into growth and distribution effect.

The socio-economic approach introduces the realities of capabilities, vulnerabilities and sustainability of livelihoods. Desirable outcomes, therefore, go beyond income gains to welfare gains, reduced vulnerability, improved food security and sustainable use of natural resources, which have strong equity considerations and therefore work up from micro to macro.

Inequality Concerns in the analysis of globalization

Divergence is a feature of current globalization. The inherent equity consideration in the implementation of globalization policies lies in:

1. Use of rules that benefit the powerful countries at the expense of he weaker nations or regions, leading to unequal rate of economic and social development which shows as deepening poverty of developing countries compared to developed ones (Human Development Reports)
2. Its human expression which hinges on the extent to which all groups within a nation can participate in its income–growth process. Unequal participation of people and groups in the new trade opportunities created by globalization bring about the paradox of income growth and increasing poverty, of segments of the same population.

In many instances especially in agrarian economies of SSA, equality of participation, equality of response to opportunities is something that cannot be guaranteed. Therefore, while many of the earlier analysis of liberalization effects have been positive on growth, more recent studies have shown that these benefits of globalization are not automatic and are contingent on the management of other evolving phenomena, the two principal concerns being inequalities and market volatility. Thus a people-centred analysis of globalization effects cannot take a purely neoclassic view of economic growth but rather, a human development approach.

Neoclassical economic analysis assesses outcomes of policy in a gender neutral or household or per capita unit of analysis. Distributional effects are considered or emphasized. Resultant inequalities are seen as incidental rather that the cause and effect of mainstream policies. Emphasizes aggregates of growth, efficiency, income and consumption.

Heterodox (feminist) economic analysis goes beyond the unitary analysis to a disaggregated view of society to answer the question: are these outcomes equal for all groups or socio-economic groups. If not, growth and efficiency have been attained at the expense of social justice and equity or at the price of inequity. That is why the theory of inequality is central to the analysis of poverty effects of macro policies such as liberalization.

The Nature and Measurement of Gender Inequality

First let us look at the dimensions of inequality with a view to situating gender inequality. Inequality exists in a development system which concentrates opportunities. It results in distributional imbalances in development outcomes such as economic well being, paucity of it or simply lack of it. Its measurement is usually a measure of concentration, which is estimated as the Gini Coefficient – a measure of the level of evenness of distribution.

Amartya Sen identifies several faces of inequality, with gender inequality being the most profound, being that it crosscuts other forms of inequality. At the micro level therefore, gender inequality has been seen as that which mostly undermines human development. At the macro level, such inequalities transmit their effects either to enhance or to constrain growth. In the latter, where inequality enhances growth, this has been termed the capitalist accumulation effect, while in the former, it has been termed the underdevelopment effect. Either way, inequality manifests as varying levels of lack or poverty, food insecurity (hunger) being a major manifestation. Gender inequality reinforces the structural dimension of poverty which has also been a concern in the measurement of poverty.

The major types of Gender Inequality that are mostly related to economic production are:

- Employment Inequality reflected often as increased labour feminization without increased gender development or economic empowerment of women;
- Persistent wage gaps between men and women in employment;

- Ownership Inequality reflected often as resource constraints for women in the same production system;
- Household Inequality reflected as unequal power relations and underlies other derivatives such as employment inequality and political inequality.

Causes of Gender Inequality

- Gender role differentiation imposed by the traditional society on time, ability, availability of men and women to respond to changing opportunities;
- Gendered entitlement systems imposing disparities on access of men and women to common property resources, especially productive assets;
- Gendered nature of production which imposes differential outcomes not only on productivity of men and women but also on returns to production due to valuation bias on outputs of production (productive and reproductive work outputs).

How large is gender inequality

Some schools of thought maintain that gender inequality is not as much of a problem as it is made out to be. So how large is gender inequality? Ravi Kanbur's analysis of the size of gender inequality is very explicit. It posits that there is sub-distribution of gender categories since gender crosscuts all other dichotomies. Total size of inequality is therefore the overall distribution of inequality (V) within each sub group.

Eboh recognizes the presence of multiple gender categories, each one exerting a gendered impact on the development process. An estimation of the size and effect of inequality must decompose along these structural lines. First to effectively capture all facets and two to assess differences in the impact on sub-categories – a distinction or decomposition of the within and the between effects. To this extent, intra-gender power structures have been found to be significant in the analysis of gender relations in primary production systems.

Gender Analysis of Liberalization Policies

Inequality becomes a relevant force in globalization because it sets the initial position of men and women at the onset of changing policies of liberalization. This initial position of advantage or disadvantage determines the responsiveness to the opportunities that liberalization present. On the aggregate, the outcome of liberalization in a production system depends on the gender structure of the actors (producers) and their aggregate supply response. In the next section, we carry out a gender analysis of liberalization impacts in cross-country studies.

Going by its own doctrines of survival of the fittest and capitalist accumulation, liberalization itself initially reinforces inequalities although it is also believed to potentially have the ability to close gender gaps and other forms of inequalities over time. Another possible outcome of liberalization is to move an economy or nation from one kind of inequality to another. Thus different forms of inequality would impose adversities or negative impacts on lives of both men and women. Inequalities of different forms that become self-reinforcing culminate in poverty of those affected and underdevelopment in spite of broadening opportunities for growth.

Gender analysis of liberalization policies is a gendered deconstruction of *Stylized Facts* around its doctrines as it affects women and men and as it affects macroeconomic growth indicators. Impact of Gender Inequality on Economic Growth, Poverty and Food Security: Stylized Facts and Empirical Evidences around the developing world

Impact of Gender Inequality on Efficiency and Output of Agriculture

The broad macroeconomic outcomes of liberalization are as follows:

Increased commercialization, increased specialization and concentration, expanded market for output export promotion via the availability of farm surplus, industrial linkages are promoted. The effects on macroeconomic variables are enhanced price of commodities, increased farm income, increased flow of private investment, increased savings and increased paid employment.

However, feminist analysis has brought out other unintended likely and documented outcomes, which are the effects of gender inequalities in the agricultural work force:

Increased feminization of the labour force, increased wage inequality, pressure on women's time budget, pressure on household subsistence, negative impact on health and nutrition, negative environmental consequences as women and the poor exploit natural resources, net shortfalls in consumption.

Several Micro studies in agriculture largely support these outcomes. Quinsumbing et al tested the assumption of *pareto efficient* household allocation of production resources in Burkina Faso. They found that farm level output was shortchanged by up to 10 per cent due to allocative inefficiency in household productive resources – a result of differential gender entitlement, even in households where farm plots were evenly distributed between male and female farmers. The assumption of *pareto efficiency* in outcome was debunked. Thus inequality impedes agricultural growth and household food security.

In Cameroun, gender inequality in technological capability and access as shown by unequal power relations in the control of yield-enhancing technology, led to substantial income loss to poor households. Women farmers persisted in growing sorghum in spite of higher returns on rice and household poverty profiles worsened.

In Zambia, women farmers refused to intercrop female crop (beans) with male crop (maize) despite potential savings of 50 per cent on weeding labour and cost. This was related to land.

Macro studies often give contradictory evidence. Kanbur, 2002 argues that findings are mixed and "the jury is out". In spite of some empirical evidences to the effect of positive benefits of gender equality for growth, the groundswell of opinion is not strong enough. He concedes that inequality impedes growth and redistribution releases constraint on growth and production, leading to welfare gains, although more from he economic and social dimension and less on the power dimension. Nonetheless, it is shown that pro-growth policy choices are promoted when endowments and capabilities are more equally distributed between women and men. The evidence against the relative importance of power inequalities needs to be reassessed against the above evidences from Zambia and Cameroun!

Biliamoune on the other hand, supports the hypotheses that increased inequality between countries of north and the south are stronger for African women than they are for African men. While globalization has favoured certain nations at the expense of others, many African countries have lost potential markets, potential foreign capital flows and globalization has forced non-competitive producers out of the market. The results show that globalization has an ambiguous effect on gender inequality in a mixed sample of developing countries, but does have a negative effect on gender inequality in Africa. This implies that gender inequality in Africa is rising as globalization proceeds.

Impact of Liberalization on Employment

The normal stylized argument is that as liberalization progresses, income growth in the productive sector leads to massive entry of women into the labour force. However as income growth progresses, the rate of feminization changes depending on the stage of development. This is the theory of the Feminization U'. Many empirical studies support this theory but the study by Darity and Ertuk clarifies the segmented effect in different countries at different stages of economic growth.

Changes in feminization ratio can apply through two different effects:

Buffer hypothesis: women's labour is secondary due to lower labour value and so as industrialization proceeds, female labour declines

Substitution hypothesis: SAP induced substitution of female labour for male labour as a cost reducing strategy. Feminization enhances economic growth.

In the craft economy (agrarian), at the initial stage of market expansion, there is massive feminization. Feminization rate however falls with rising income due to greater competition with (male) skilled labour. Job losses result for women.

In middle income (semi-industrialized) countries, feminization rate proceeds at the same rate as income growth because net job loss tends to equate net job gains as women's education and skill increases.

In high income countries, feminization rate increases with increasing income (economic growth) because the opportunity cost of women's time is now higher than its subsistence value. You may find the phenomenon of "sit-at-home-husbands".

Recent efforts towards conceptualizing gendered impacts of openness (liberalization) through price, income and employment have verified this U-shaped feminization effect in many newly industrializing nations. Export promotion and trade liberalization have led to massive movement of women into the labour force. Whether this is good for women or good for growth has also been a subject of further analysis. Ertuk and Darity showed that the new global division of labour between the North and the South could thwart economic benefits normally associated with trade liberalization.

In one likely effect, female unpaid labour in the household subsidizes the reproduction of labour. This is a cost on women, which diminishes and passes to the market as higher cost of household welfare and food security as labour force feminization proceeds. On the other hand, wage discrimination against women in the paid employment is a net gain on growth.

Floro examines the same effect in agricultural households in the Philippines and notes substantial substitution of labour between men and women as agricultural export increases, putting pressure on subsistence and welfare. In Nigeria, the substitution effect on household welfare and nutrition was evident during the structural adjustment programme.

Impact of liberalization policies on value of labour

The reproduction of labour is normally subsidized by women via domestic roles of nurturing, welfare provisioning etc. As income growth proceeds and labour force is feminized, cost of reproducing labour becomes positive as industrialization proceeds. In a craft economy, the burden of care falls on the market and because of infrastructural deficiencies and underdeveloped goods and services markets, the value of labour increases as labour quality is compromised leading to inefficiency in production, declining output, declining income, high food prices and food insecurity.

In an industrialized economy, the cost of labour is minimized due to efficient goods and services market. Thus liberalization may leave the value of labour unchanged.

Effect on composition of output

With industrialization, composition of output changes from primary products to industrial goods and export-led industrialization. Female labour has been seen to subsidized production of industrial goods as unionization breaks down (sweatshop effects)., Economies change production to industrial goods at the expense of women's sweat. The phenomenon has been noted in Export Processing Zones (EPZ) and in Commodity Value Chains (Kenya), although they have been shown to provide great opportunities for women to benefit from globalization, with short-term benefits for income and household poverty.

Impact on Investment and Savings

Where the substitution effect is higher, labour force feminization is efficient. Investment increases due to the minimal cost incentive. In a cross country analysis, empirical data not only shows that GDP growth is positively related to gender wage inequality but that part of the impact is transmitted through its positive effect on investment.

During liberalization such as SAP, women's work burden increases due to cut down in governments social spending. The share of household goods that substitutes for market defeminization, leading to desaving. But the higher the household provisioning, the higher the savings. Poverty management and food security occurs at the expense of women.

Higher subsistence leads to higher savings and declining poverty. On the other hand, higher market income leads to increasing opportunity cost of female labour, decreasing subsistence and decreasing savings. Liberalization may affect savings negatively.

Increased Paid Work Leads to Economic Empowerment for Women

Again, findings are highly mixed and contextual. While in agrarian economies, the economic opportunities are enlarged for women, in other subsistence and newly industrializing economies, there are often less clement effects on women's position in the workforce.

Changes in the conditions of (paid) work due to more informalisation of work, less unionization have led to depressed wages and the sweatshop syndrome. Wage differentials are reinforced. In developed countries, there is evidence of drop in gender wage gaps, given enhanced competition between men and women skilled workers in the work place. In developing countries, wage gaps are exacerbated due to low skill of new female market entrants, high supply of labour and low bargaining power of the unskilled vis a vis the skilled. Work normalization and feminization of work rather is the norm to the advantage (capitalist accumulation) of the market economy.

Occupational segregation also persists due to skill differentiation and because skill upgrading is costly for women. Net job creation effects may be unclear. Net job creation increases for women irrespective of skill category. But net job destruction is also rising in high technology-intensive environment. Volatility of employment and income is therefore on the rise for female labour.

For large industrial firms, the above picture is clearer. Extensive studies in South East Asia and Latin America have been more amenable to the analysis. For agriculture, the informal nature of employment, smaller firms, and high-unpaid work burden makes t more difficult to show the effects from cross-sectional data because of preponderance of unskilled workers. Net job creation may appear to be higher. However income effect is usually negative while wage gap effects are reinforced.

Trend analysis of agricultural data shows the structural effect more clearly as the sector is being transformed in size and technology and income. Income effect becomes lower and wage gap between skilled and unskilled jobs increases. In developed economies, the reverse is usually the case. So in agrarian economies, the stylized feminization 'U' shows up only at the later stages of structural transformation.

Impact on Agricultural Growth via Small Farmer Commercialization (SFC)

Increased commercialization of agricultural production is seen as a positive response to adjustment policies. The optimal response is expected to 'hasten' the process of structural transformation. Therefore, any constraints to supply-response would naturally show down the rate of commercialization. Conversely, optimal trends in the factors that are critical determinants of supply-response should be positively related to the pace of commercialization. In the sense that such factors or the availability of factors of production vary by gender, it is expected that rate of commercialization will vary by gender.

An identification of the critical determinants of supply response and differentials in these parameter estimates for male and female farmers (degree and dimensions of gender inequality in assets ownership, capabilities and accessibility) will lead to different set of outcomes than projected by policies. Priorities for bridging gender gaps are called for in hastening the process of agricultural commercialization and growth.

Some of the determinants of commercialization have been analysed conceptually and empirically and gender differentials are estimated. The critical determinants are largely related to the macroeconomic growth variables, albeit at the micro level.

Level of investment which may derive from capital accumulation from women in terms of inelastic labour demand is shown to be important. Gender inequality in labour utilization may enhance commercialization for men. On the aggregate, the gross investment is likely to be shortchanged. Credit becomes crucial in an analysis of investment behaviour of commercial farmers. Therefore, what credit volume is available to male and female farmers is a determinant of commercialization.

Land Asset is another critical factor, and consistent with the 'financial hierarchy hypothesis', bigger financial (asset) base predisposes to higher access to credit and better supply-response or rate of commercialization, inter alia. Land is the most important asset base in small-scale agriculture. To the extent that access of farmers to land vary significantly by gender this variable is crucial to the analysis of commercialization.

Cropping pattern as a derivative of land size is also important. Land, on its own, may be insufficient, rather the optimal land use within the framework, of commercialization is more important. In this sense are the allocation of land to commercial and non-commercial production and the ratio of traded to non-traded crops on male and female farms become critical.

Inequalities in these initial positions during Small Farmer Commercialization (SFC) have been seen to negate the expected aggregate outcomes from SFC. Food security objectives are still paramount with small farmers. In a study of several African countries, the following patterns were observed:

SFC is part of a more comprehensive change in rural livelihoods which include off-farm activities especially wage labour for women. Evidence is also given that it is part of a long-term transition from traditional peasant life to new rural lifestyles (globalization effect).

A major finding in the process of SFC is short-term food shortages at the household level due to spontaneous development of a market economy. Intensification of staple crop production does not translate into food availability, rather, enhanced income from a variety of livelihood options buttress food security in the medium term. Food price hikes are common in a SFC environment.

Diversification, rather than specialization was therefore observed. More plot fragmentation took place rather than land amalgamation. Therefore, the food security objective is stronger in Africa's SFC than themercerization objective.

Impact of Socio-cultural Variables

In the sense that the socio-cultural variables exert particular power structures between men and women in rural communities, they are all expected to produce differential gender effects on investment behaviour and level of farm commercialization. Variables like literacy will enhance participation in the market economy via greater access to information and higher technological learning. The age and gender of the farmer are important power variables in a rural community. Even among women, age crosscuts with gender to exert particular influences. An older woman in a farm household has the status of a man among other women while her position among the wives may confer her privileges such as greater access to land and other communal property resources. The nature of employment is also important.

Being primarily engaged in a particular crop production confers social capital, which is found in commodity-based social institutions, like farmers associations or cooperatives. Other (part-time) growers may lack such influence while on the other hand, having another source of income apart from rice farming could be a boost to investment capacity. These are contentious issues and contextual in manifestation and therefore need to be empirically validated.

The foregoing has shown the extent that liberalization policies impact of different people in different sectors of an economy. While liberalization policies will be gender-neutral in their indication, they are hardly gender-neutral in their outcomes. It takes an analysis of such outcomes by gender categories to see the differential impact. It has also been shown that such outcomes of gender inequality do impact on the overall economy, mostly in negative ways. One, disincentives are set up, productivity is shortchanged, discrimination in wages either fuel growth of profit-oriented firms but against unethical operating standards and/or creates disenfranchised population of workers. Inequalities in liberalization effects affect productivity and output especially in agricultural systems, thereby worsening food security. It takes a gender analysis framework to unpack these effects. Gender analysis instruments are an important tool for interrogating reform policies in order to provide a framework to mainstream gender in the workings of policies themselves.

9

Food Security and Rural Development

AGRICULTURE AND RURAL DEVELOPMENT

From a nation dependent on food imports to feed its population, India today is not only self-sufficient in grain production, but also has a substantial reserve. The progress made in agriculture during the last four decades has been one of the biggest success stories of independent India. Agriculture and allied activities constitute the single largest contributor (almost 33 per cent) to the Gross Domestic Product. About two-thirds of the work force in the country depends on agriculture as a means of livelihood.

Despite these impressive gains, India, at present, finds itself in the midst of a paradoxical situation: On the one hand there are record food grain stocks standing at an all-time high (62 million tonnes against an annual requirement of around 20 million tonnes for ensuring food security), and on the other hand, over 200 million of India's population is underfed, and millions are undernourished. The challenge is to bridge this gap. In a scenario of shrinking land and depleting water resources, the challenge of the new millennium is to increase biological yields to feed the ever-growing population without destroying the ecological foundation. It is thus important–not to package this challenge as a demand or imposition of society on farmers, for which farmers would bear the cost, but as a necessity and methodology to also sustain their welfare and incomes.

India has the potential to meet these challenges. This potential can be realised through policy and infrastructure support from the government and by strengthening proactive synergies among the various sectors that play influential roles in the field of agriculture and rural development. The National Policy on Agriculture seeks to actualize the vast untapped growth potential of Indian agriculture, strengthen rural infrastructure to support faster agricultural development, promote value addition, accelerate the growth of agro-business, create employment in rural areas, secure a fair standard of living for the farmers and agricultural workers and their families, discourage migration to urban areas and face the challenges arising out of economic liberalization and globalization. The following are suggestions towards achieving sustainable agriculture, rural development and food security. They provide a reference for actions to be undertaken by the various stakeholders in this area.

PROMOTE SUSTAINABLE AGRICULTURE

Agriculture is directly linked to very many facets of sustainable development, including poverty eradication, sustainable consumption and production, management of natural resources, energy, freshwater, health, education, trade and market access, as well as technology transfer and capacity building. Agriculture is an integral part of the general development system, serving the system as a whole, and being served by it. If the effects of other sections of the development systems reduce sustainability, then sustainability of agriculture is also affected. A sustainable system should be resilient, and able to withstand shocks and failures of parts of its systems without the whole collapsing, and without small shocks leading to a spiral of unsustainability.

Agriculture centres on integrated use of natural resources such as soil, water, climate and biological diversity. The integration of agriculture with other aspects of land management and ecosystem conservation is essential in order to promote both environmental sustainability and agricultural production:

- Natural resources have to be accessible to the poor, which in the farm sector means secure rights to land, water and genetic resources. For this, there is a need to develop public-private partnerships.
- Policies for land and water resource management, biodiversity protection, infrastructure investment, strategy on institutional market reforms, reduction of tariffs and phasing out of possible commodity control are integral to achieving sustainable agriculture. ?Sustainability should be seen in the context of different agro-climatic zones as well of as the country as a whole. Suitable technologies should be developed and indicators for sustainable agriculture should be identified for both.
- It is essential to provide funding for integrated rural development plans, programmes and strategies, at national and regional levels, with particular emphasis on investment in economic and social infrastructure in rural areas, enterprise development, human resource development, and capacity building for local governance.
- An efficient credit policy regime with the required rural banking and credit system will play a major role in the future. The government should encourage investment in vital agriculture infrastructure, credit linkages, and use of new and appropriate techniques towards this end.
- More than 70 per cent of the country's agriculture is under small and marginal farmers with limited resources. A nationwide crop insurance scheme will provide such farmers the needed confidence to invest and gain from technological advances in agriculture.
- The focus on accelerated food grains production on a sustainable basis and free trade in grains, as well as on rural employment opportunities will lead to faster economic growth and give purchasing power to the people, which in turn would help increase household food security.

- Concerted efforts should be made at national, regional and local levels to pool, distil and evaluate traditional practices, knowledge and wisdom and to harness them for sustainable agricultural growth.
- It should be recognized that information is a critical input for agricultural development. It is as important as other key inputs including credit, seeds, nutrients and water. Information can be efficiently converted into economically rewarding opportunities.
- It is critical to recognize that the challenge to world agriculture is both technological (requiring the development of new, high productivity, environmentally sustainable production systems), and political (requiring policies that do not discriminate against rural areas in general, and agriculture in particular).

Equitable Distribution and Access

A positive right to life would imply that the State provides to each and every person, adequate food and other basic necessities, and that it ensures a healthy environment, so that people may live and grow in dignity. The most important challenge in the 1980s and 1990s was physical access to food. In this millennium, the challenge is economic and ecological access to food.

- A transition from chemical and machinery-intensive technologies to ecological farming technologies is required towards providing sustained physical access to food.
- Environmental access involves on the one hand, attention to soil health care, water harvesting and management, conservation of forests and biodiversity, and on the other hand to sanitation, environmental hygiene, primary health care and primary education.
- Emphasis on economic access underlines the need for promoting sustainable livelihoods through multiple income-earning opportunities.

- It is important to increase food availability in areas where it is produced, thus reducing transport costs and excessive dependence on international markets.
- Shift from existing expensive, inefficient and ineffective institutional arrangements, to decentralized management systems of food storage and distribution will improve delivery, reduce handling and transport costs, and reduce corruption, thereby bringing down the issue price substantially.
- Procurement of grain can be decentralized through creation of food grain banks in each village/block of the district, from where people can get subsidized foodgrains (including locally grown coarse cereals) through food coupons.

Secure Food Security for All

Food security is a physical, environmental, economic and social issue. It involves not just production, but access; not just output but process; not just technology but policy; not just global balance but also national conditions; not just national figures but household realities; not just rural but urban consumption; and not just quantity of food but also quality.

- The concept of food security should be broadened to make it holistic so as to mean "every individual has the physical, economic and environmental access to a balanced diet that includes the necessary macro and micro nutrients and safe drinking water, sanitation, environmental hygiene, primary health care and education so as to lead a healthy and productive life."
- Issues of food security are part of a bigger whole. Sustainable land and water management must be seen as directly linked to food security. Population growth, environmental sustainability, poverty reduction, agricultural production, distribution, marketing, credit and many other factors also need to be recognized as part of this whole.
- The major challenge is to produce additional food while conserving depleting natural resources. It is also

to provide physical, economic and ecological access to food and nutrition security at the household level.

- Food security must focus on a diversified food basket, not food grains alone. Broad-based food security systems are not dependent on two-three species but on over 100 species that are underutilized. This will also prevent locally adapted grains from becoming extinct.
- Nutrition security must be given integrated attention by emphasizing horticulture, animal husbandry, fishery, millets, pulses and several other resources for which India is traditionally known. There is need for investment in science and technology that will promote diversification.
- Food security must not be based on market, but rather on self-reliance and sufficiency. The approach should be one of moving from food security to food sovereignty.
- The elimination of hunger and malnutrition is not just a food problem. It is linked to poverty and population growth. Rising food output is essential but so are the slowing of population growth and maintaining the ecological balance.
- Food banks at grass root levels should be well maintained.

Capacity Building Mechanisms

In the 21st century, it is increasingly necessary, and increasingly feasible to take a whole systems approach to organized, positive change in rural places. For extension, that means helping farming people toward sustainably increasing productivity–particularly in the small-mixed farming systems in rain-fed areas, in upland areas, and in other places which have been neglected. It also demands measuring success in terms of the consumption of rural people, as well as of their production. And that, in turn, will require agricultural extension systems which help farm men and women organize themselves in ways which empower them–to lead agricultural extension and to exert enough power and influence over agricultural research systems so that they generate useful, practical information which fits the needs and interests of those farming people.

- Agricultural extension must focus on increasing production and productivity of food and fiber in an economically and environmentally sustainable way. It must be done in a way which does not destroy rural livelihoods and rural communities.
- Extension activities should promote more comprehensive rural education and extension programmes directed particularly at rural poor, with major emphasis on efforts to reduce illiteracy, particularly among women and girls.
- It is important to organize education, extension and information, and skill empowerment on the basis of intensification, diversification and value addition of farming systems.
- Partnerships aimed at strengthening the knowledge base, and improving the dissemination of information, such as farm-to-farm technical assistance programmes, can help strengthen agri-extension. Public-private partnerships could be envisaged in basic sustainable agricultural techniques.
- Sustainable agriculture demonstration plots should be set up in research stations, demonstration centres, seed production centres, farmers training centres etc., of government, cooperative and non-government agencies.
- The system should support a new agricultural extension system that could meet the needs of information-hungry farmers, especially educated youth and women engaged in farming, and would empower them with new techniques and skills that foster sustainable agriculture.

Promote Awareness and Education Activities

National policies and planning should recognize that public awareness can play an important role in establishing a firm basis for sustainable agricultural resource conservation and use. Public awareness should be considered in the development of all national programme activities.

- National strategies should identify objectives and strategies for public awareness, define target audiences, partners and tools for public outreach.
- Governments should recognize and encourage the work of NGOs in raising public awareness.
- Short-term and long-term courses specializing in sustainable agriculture should be run in agricultural universities. A separate curriculum for this subject should be developed for the regular graduate and postgraduate courses also.
- There is generally public apprehension about the possible consequences of agricultural biotechnology. It is important to organize public information programmes and public discussions that would help share relevant work of agricultural and biological scientists with the public.

Application of Research, Science and Technology

Scientific and technological development is not enough in itself. It is critical to make sure that sustainability and poverty reduction remain the guiding principles, and that we use our resources, harness our intellect, and direct our knowledge to benefit the poor, the hungry and the marginalized.

- The impact of agricultural research is decisive. Such research should, wherever possible, be coupled with on- farm activities in order that the context and purpose of the work are fully appreciated. Research should assist in the monitoring, evaluation, and improvement of on- farm efforts.

- Research should be undertaken in a participatory and collaborative manner to foster interaction and cooperation between rural people and research institutions. Other institutions must be involved appropriately whenever necessary.
- A comprehensive area-specific database of natural resources should be developed and made available for agriculture planning, implementation, research and extension. Existing data and information should be assembled, verified, and put in a usable and easily accessible form.
- Well-designed information technology packages should be developed, that could help serve as a market information network; weather, pest and disease monitoring system; and could be a storehouse of various current farming technologies and practices.
- Modern information technology should be used to reach the unreached. Educated youth must be attracted to and retained in farming through spreading science-based precision farming techniques, which are intellectually stimulating and economically rewarding.
- Training and capacity building should be undertaken in areas such as taxonomy, population biology, ethnobotany, and eco-regional and agro- ecological surveying. Specific research priorities need to be determined separately for each region.
- Additional research is necessary to develop drought tolerant, pest and disease-resistant crops, biological pest management, nitrogen fixation, more effective use of locally available organic materials, inter-cropping systems, and perennial crops, including agroforestry.
- Farming systems need to be designed so as to achieve the triple goals of more food, more income and more livelihoods per hectare of land. For this, it would be fruitful to harness the tools of eco-technologies resulting from a blend of traditional knowledge with frontier technologies. Such tools include biotechnology,

information and communication technology, GIS mapping, space technology, renewable energy technologies (solar, wind, biomass, biogas), and management and marketing technologies.

- The revolution in biotechnology has both promises and problems. Biotechnology should be judiciously used so as to support the mission of environmental protection, poverty reduction and food security. But the adoption of every new technology must be accompanied by a precautionary package. In the case of biotechnology, biosafety and biosurviellance must be considered as important factors.
- Food should originate from efficient and environmentally benign production technologies that conserve and enhance the natural resource base of crop and animal husbandry, forestry, inland and marine fisheries.

The Value of Agricultural Biodiversity

While a small number of species provides a large proportion of global food needs, hundreds of other species are utilized at a local level, either through cultivation or harvesting. These under-utilized species contribute substantially to household food and livelihood security. They are often managed or harvested by women. Knowledge concerning the uses and management of these species is likewise often localized and specialized. Many under-utilized plants have potential for more widespread use, and their promotion could contribute to food security, agricultural diversification, and income generation, particularly in areas where the cultivation of major crops is economically marginal.

- It is vital to recognize the intrinsic value of biological diversity and of its ecological, social, economic, scientific, educational, cultural, and aesthetic importance. This diversity is being lost in the fields and other ecosystems of virtually all countries.
- High priority needs to be given to safeguarding as much existing unique and valuable diversity as

possible in *ex-situ* collections of plant genetic resources for food and agriculture, and also through *in-situ* conservation in their natural habitats.

- Goal-oriented, economically efficient and sustainable system of *in-situ* and *ex-situ* conservation needs to be developed.
- Cooperation among national programmes and international institutions to sustain *in-situ* and *ex-situ* conservation efforts needs to be developed and strengthened. It must be recognized that states have sovereign rights over their own plant and animal genetic resources for food and agriculture.
- A time-bound programme to list, catalogue and classify the country's vast agro-biodiversity with special focus on conserving indigenous breeds and species, must be initiated.
- Agriculture diversity registers should be formed for local and regional varieties. Agriculture policies should be formulated based on these registers.

Relook at Agriculture and Related Policies

It is critical to develop and strengthen appropriate policies and legislative measures to create an enabling environment for sustainable agriculture and rural development. Such an environment would promote access by the poor to land, water resources and other agricultural inputs; land tenure modifications that recognize and protect indigenous and common property resource management systems, and also build local capacities for better management of natural resources.

- Governments should consider, and as appropriate, adopt policies in extension, training, pricing, input distribution, infrastructure development, credit and taxation, which serve as incentives for crop diversification and the creation of markets for biodiverse food crops, including standards for labeling of foods, which allow the highlighting of use of non-standard crop varieties.

- Policies should stress on land reforms, input and output pricing, investments in irrigation, infrastructure and insurance, legislation for biodiversity, geographic appellation, varietal protection and farmers' rights.
- Well-defined and enforceable land rights, legal security of tenure and equal access to land, water and other natural and biological resources, need to be assured, in particular for indigenous communities, women and disadvantaged people living in poverty.
- It must be recognized that states have sovereign rights over their plant genetic resources for food and agriculture, while also confirming our common and individual responsibilities in respect of these resources.
- Coordination is needed to provide national programmes with information on these issues and to assess the impact of international developments in these fields on the conservation and exchange of plant genetic resources, and to incorporate new research developments, as appropriate, into national systems and practices.
- Effective regulatory mechanisms and safeguards need to be universally installed so that the impacts of biotechnologies are both productive and benign.
- Development of agro-ecology-relevant technologies based on an understanding of local agriculture and resource management practices need to be supported and promoted.
- Chemicals and pesticides banned in developed countries should not be dumped into developing countries in the name of liberalization, globalization and industrialization.
- When patenting a variety or item of research, the area of origin should get due credit and benefit of the patent.
- Before releasing a new variety, including genetically modified varieties in the market, the following

parameters should be considered for a variety of agro-climatic zones: Impact on soil productivity; hazardous residual effect; health hazards; adverse effects on other crops; adverse effects on other agricultural practices; threat to the indigenous varieties; impact on other professions; and impact on flora and fauna.

Create Favourable Economic Climate

Agriculture has become a relatively unrewarding profession due to generally unfavourable price regimes and low value addition, causing abandoning of farming and increasing migration from rural areas. The situation is likely to be exacerbated further in the wake of integration of agricultural trade into the global system, unless immediate corrective measures are taken. A favourable economic environment and supportive public management system are the key pillars for the promotion of sustainable agriculture.

- Capital inadequacy, lack of infrastructural support and demand side constraints such as controls on movement, storage and sale of agricultural products, etc. affect the economic viability of agriculture sector. These issues need urgent attention.
- Increasing capital formation and farmer's own investments by removal of distortions in the incentive regime for agriculture, improving the terms of trade with manufacturing sectors and bringing about external and domestic market reforms, backed by rationalization of domestic tax structure, will help to create a favourable economic environment for agriculture.
- Agriculture-dependent countries like India should have a certain degree of autonomy and flexibility in determining the domestic agricultural policies so as to improve productivity, enhance income levels, reduce vulnerability to market fluctuations, ensuring stability of prices, etc.
- Creating a level playing field in the global marketplace will provide the necessary incentives and leeway for

farmers, especially the small ones to adopt environmentally friendly farming practices, and help in avoiding, their desperate acts of survival at the expense of sustainable development, due to distortions in international trade.

- In the context of globalization of the food market, farmers in developing countries need to improve access to their own local markets. A 'market- plus' rather than a pure market approach for the agricultural sector would be effective in addressing the wide range of issues associated with production, pricing, food distribution and access.
- At international forums, developing countries like India should have the necessary flexibility to pursue legitimate non-trade concerns with regard to international markets. Such countries should seek reforms in global agricultural policies relating to price, subsidies, trade and technology transfer, and respect for IPRs of local communities.
- International agreements should allow room for the domestic agricultural sector to meet challenges of maintaining the livelihood of the large agriculture-dependent population, and production of sufficient food to meet domestic needs.
- Agricultural food processing industries with international health standards should be promoted locally.
- Biomass based entrepreneurship should be promoted to generate wealth at the rural level.
- Civil society groups should use benchmarks and sustainability indicators to monitor performance of food processors, retailers and food service companies, especially regarding fair terms of trade.

Ensure Participation at All Levels

National programmes are increasingly confronted with policy, legal and institutional issues related to ownership, intellectual property rights, exchange, transfer and trade in agricultural resources. Participatory processes and involvement of various stakeholders can help find answers to such concerns. Towards this it is crucial to develop a framework for a unified national programme to enhance the diverse efforts within the country to tackle such issues.

- Programmes and policies for food and agriculture activities should involve public and private institutions and companies, non-governmental organizations, communities and individuals from the agriculture, environment and development sectors.
- The role of indigenous and local communities in conserving, collecting, improving and sustainably using plant genetic resources for food and agriculture must be acknowledged. Local, community-level initiatives and participation in proposing programmes need to be encouraged.
- Gender concerns in agriculture need to be mainstreamed. Appropriate structural, functional and institutional measures to empower women and build their capabilities and improve their access to inputs, technologies and other farming resources need to be strengthened.
- The need for equitable sharing of benefits arising from the use of traditional knowledge, innovations and practices relevant to the use and conservation of plant genetic resources for food and agriculture must be recognized.
- Nutrition security must be placed high on the agenda for development plans and programmes at all levels– village, block, district and state. Institutions that will design and monitor locality-specific interventions must be developed.

10

The Future of Food

INDIAN AGRICULTURAL GLOBALISATION

In India and elsewhere, globalisation of agriculture, under the triple pressure of World Trade Organisation (WTO) rules, World Bank-IMF conditionalities and structural adjustment programmes, and economic and political interests of the national elite is destroying the land and those who labour on it, small peasants and landless workers. The very viability of a national agriculture and food security sys- tem is being ruptured as high cost, capital intensive, corporate farming displaces small farmer centred agriculture, and trade replaces food rights of the poor as a policy priority.

A reading of the new agricultural policy brings out three core areas of policy shift that are destroying the fragile fabric of ecological security, livelihood security and food security, creating ecological devastation and deepening hunger and poverty. These are:

1. Changes in property rights to natural resources—land, water and biodiversity (including seeds) on which agricultural production rests;
2. Changes in technology, especially new genetically engineered crops and seeds (GMOs) which promote use of agrichemicals, increase costs of production for farmers and corpor- ate control over agriculture policies and practices;

3. Changes in trade regimes, privileging and liberalizing exports and imports, thus undermining national food security, farmers' livelihood and food rights of the poor.

Changes in property rights to natural resources: land, water, biodiversity Trade "liberalisation" policies are leading to the alienation of land, water and biodiversity from peasant communities and the concentration of their ownership in the hands of corporations. New private property rules created by W.T.O. under TRIPS (Trade Related Intellectual Property Rights) and GATS (General Agreement on Trade in Services) are leading to the enclosure of biological resources and water which have hitherto been common property.

CONCENTRATION OF LAND OWNERSHIP

At the heart of colonial policies for extraction of revenues from agriculture was the introduction of the "zamindari system". Zamindari abolition was one of the most important steps taken in independent India and the most significant instrument of social justice was the land reform legislations in different states to ensure equitable entitlement to land and to prevent concentration of land ownership. Land reforms such as operation Barga which put land back in the hands of the tiller were also introduced in Bengal. However, during the last ten years globalisation and economic reforms in agriculture have in effect meant an undoing of the earlier reform process guided by values of social justice and equitable distribution of resources. While the positive protections afforded to small farmers and poor consumers and to self-reliance in food for the country have been removed, the reform' package has increased the tendency of centralised control over agriculture.

The main argument used for the industrialisation of food and corporatisation of agriculture is the low productivity of the small farmer. But in terms of food and nutrition productivity per acre, in terms of efficiency in water use, in terms of creation of livelihoods small farms are more productivity than large ones. Even the World Development

Report (WDR) has accepted this fact. Even biologically, small diverse farms have higher productivity than large monoculture farms as long as multiple yields are taken into account. FAO data on total farm output confirms this.

In India, a 0–5 acre farm had a productivity of Rs. 735/acre while a 35 acre farm had a productivity of Rs. 346/acre. The state of Bengal was showing the highest rate of growth of 6.5 per cent for agriculture as a result of land reform, while the rate of growth for India was a mere 3 per cent [9]. Under World Bank Structural Adjustment pressure, combined with the arrival of a new breed of absentee landlords or "zamindars" industrialists, agribusiness corporations, speculative investors—land reform laws in every state are being undone, alienating the land from small producers and cultivators, swelling the ranks of the landless, the dispossessed, the unemployed. For instance, Karnataka has amended which undoes the radical reforms that made the tillers the owners of land and prevented non-agriculturalists from becoming absentee landowners.

Described as 'predatory capitalism and legalised land grab', these amendments reintroduce land leasing, allow non-agriculturalists and industrialists to own land, and remove land ceiling for aquaculture, horticulture, floriculture and housing industry. The government of Maharashtra has relaxed restrictions on conversion of agricultural lands to non-agricultural land. The agricultural Land Ceiling Act has been amended to permit large land holdings leading to skyrocketing land prices.

Water Privatisation, Water Monopolies

Trade liberalisation is also leading to the privatisation of water and creation of water monopolies. Water is being reduced to a commodity, owned and traded by water giants—Suez, Vivendi, Bechtel, Thames-RWE. The World Bank policy paper on liberalisation of agriculture recommends the creation of 'markets in tradable water rights'. It argues that 'if rights to the delivery of water can be freely bought and sold, farmers with new crops or in new areas will be able to obtain water pro- vided they are willing to pay more than its value to existing users, and established users wiil take account of its sale value in deciding on what and how much to pro- duce'. This institution of tradable water rights is a guarantee for diverting water from small farmers to large corporate super farms. In the logic of the market, tradable rights have a tendency to be sold to the highest bidder and hence lead to water-power linked to concentration of wealth, and to over-exploitation and misuse of water, since those who deplete water resources do not have to suffer the consequences of water scarcity.

Besides, aggravating the already severe ecological crisis in water resources, tradable water rights will destroy the social fabric of rural communities and create discord and disintegration. The social breakdown in Somalia can be traced, in part, to the privatisation of water rights according to the World Bank policy. Tradable water rights are based on the assumption that no ecological or social limits should be placed on water use. Such use without limits leads to abuse. For instance, the new Agriculture Policy of Karnataka talks of a shift from "top down to bottom up" approach. What it implies, however, is that the control over water resources will move upwards from small and marginal farmers to large corporations and agribusiness interests who can buy up the "water equity shares" of "water users associations" and establish monopoly control on water.

The massive $200 billion project of River Linking will also rob rural communities of their riparian water rights. The project, the government claims, is justified on grounds of increased food security through increased irrigation. However, higher water use does not translate into higher food production or nutritional security. Green Revolution crops use 5 to 10 times more water than native varieties. Paddy and Sugarcane use 5 to 10 times more than millets, which provide higher nutrition. Food production could be increased five fold without increasing irrigation by shifting to water prudent but nutrition rich millets.

The sale of Ganga water to the French company Ondeo Degremont (subsidiary of Suez Lyonnaise des Eaux Water Division—the water giant of the world) for the Sonia Vihar water treatment plant in Delhi exemplifies the way privatisation of water is being pursued. 635 million liters a day of water would be made available from the Upper Ganga Canal, one of the oldest canals in Western U.P. While Delhi residents will benefit and the company expects to make Rs. 100 million (10 crore) per annum, mainly based on government guarantees as in the Enron/Dabhol project, farmers of Western U.P. will loose Rs. 2 billion annually, the total "investment" by the company. This project will obviously affect the agricultural output and the food security of the region where the canal had been irrigating for more than one century. The farmers are understandably agitated and the people staying in the region have opposed the project, asserting that the "Ganga is not for sale".

Patents on Life and Seed Monopolies

Seeds and biodiversity, which have been the common property of farmers and local communities, are being transformed into private property of a handful of corporations—Monsanto, Syngenta, Duport, Dow, Bayer. These chemical giants now control seeds as well as medicine. India, in fact, is losing its biological and genetic resources, the basic wealth of our country, and the biodiversity based knowledge of our local communities mainly through biopiracy. The piracy and patenting of our rich biodiversity by the MNCs, institutions and individuals especially from the Western world is inflicting great injury to the natural resources of our land and people. If this is not arrested by appropriate legislation and policy measures, there would be incalculable damage done to the nation and the coming generations.

India is a signatory of the two international treaties that have an impact on the nation's biodiversity wealth and on peoples' rights to use it and conserve it responsibly. These are the Convention on Biological Diversity (CBD) and the Trade Related Intellectual Property Rights (TRIPs) Agreement under the WTO. India on behalf of a group of countries including Brazil, Bolivia, Cuba, Dominican Republic, Ecuador, Thailand, Peru and Venezuela has submitted a paper to the TRIPs Council of WTO, on "The Relationship Between the TRIPs agreement and the Convention on Biological Diversity and the Protection of Traditional Knowledge", calling for amendments in TRIPs to harmonise it with the CBD.

However, at the domestic level, India has harmonised its Biodiversity laws to be consistent with the unamended TRIPs agreement. The Government of India has enacted new legislations to implement these international obligations which would far reaching implications for the lives and livelihoods of millions of Indian people. The new corporate-driven intellectual property rights regimes (IPRs), especially the Trade Related Intellectual Property Rights Agreement of WTO is leading to seed monopolies and biopiracy. Seeds have been evolved by nature and farmers over millennia.

This collective, cumulate heritage is now either being destroyed by introduction of monocultures of non-renewable seeds or being hijacked by global corporations through patents and biopiracy. When a seed is patented or covered by breeders' rights, farmers can no longer save or exchange seed freely. Seed saving and seed exchange in defined as a "theft" in intellectual property law. The Indian legislations that have an impact on biodiversity and people's rights are: Patent Act, 2002; Protection of Plant Variety Protection and Farmers' Rights Act, 2001; and the Biological Diversity Act, 2002. The country has amended its Patent Act 1970 for the second time, in May 2002, since TRIPS came into force. The first amendment was to introduce exclusive marketing rights and mail box arrangement to implement Art. of TRIPS.

There are two amendments in the definition of what is not an invention that has opened the floodgates of patenting of genetically engineered seed. First, in Section of the Patent Act, the word "plants" have been omitted. According to Section the following is not an invention: 'Any process for the medical, surgical, creative, prophylactic or other treatment of human beings or any process for a similar treatment of animals or plants or render them free of disease or to increase their economic value or that of their products.' The omission of "plants" from this section implies that a method or process modification of a plant can now be counted as an invention and can hence be patented. Thus the method of producing Bt. cotton by introducing genes of a bacterium Bacillus thurengensis in cotton to pro- duce toxins to kill the bollworm can now be covered by the exclusive rights associated with patents. In other words, Monsanto can now have Bt. cotton patents in India.

The second amendment has also added a new section. This allows production or propagation of genetically engineered plants to be counted as an invention, and hence patentable. The section excludes as inventions "plants and animals...including seeds, varieties and species and essentially biological processes for production or propagation of plants and animals". However, the emergence of new biotechnologies is often used to define production of plants and animals through genetic engineering as not being essentially biological.

The Biodiversity Act passed by the Parliament in December 2002 was to implement India's obligations under the Convention on Biodiversity Conservation (CBD) of the United Nations. The breakthrough achieved in the CBD was:

1. The need to conserve biodiversity;
2. Recognition of community rights and indigenous culture and knowledge;
3. The prevention of adverse impact on biodiversity by mega development projects and by the release of genetically engineered organisms (GMOs), and
4. The potential to prevent biopiracy the theft and patenting of indigenous knowledge related to biodiversity.

Earlier drafts of the Bill had all four components. However, the draft passed by Parliament has no conservation element, no recognition of community rights to Biodiversity (even though this is necessitated by the Panchayati Raj Amendments of the Constitution and the extension to Schedule Areas) no regulation of adverse impacts. Given that the Environment Ministry, through the Genetic Engineering Approval Committee has been rushing ahead to commercialise genetically engineered crops like Bt. cotton in spite of known risks to biodiversity and in spite of exaggerated claims of benefits, it is clear that pressures were put on the ministry to drop clauses on regulating adverse impact in earlier drafts. The act is reduced to an act to facilitate access to biodiversity and indigenous knowledge for intellectual property rights. It facilitates and legalises biopiracy, instead of preventing it.

All that biopirates will need is a cosy relationship with the National Biodiversity Authority. In the case of indigenous seeds and agricultural crops, the Biological Diversity Act has allowed full scope for biopiracy. Section on intellectual property has an exemption:"The provisions of this section shall not apply to any person making an application for any rights under any law relating to protection of plant varieties enacted by Parliament." Laws protecting plant varieties include the Patent and the Plant Variety Protection Act. Exemption in the Biological Diversity Act in effect says that companies can take varieties farmers have evolved over millennia with unique traits of aroma as in basmati, salinity resistance, drought resistance, flood resistance, and patent the traits and qualities which are a result of farmers breeding. A global movement is calling for a ban on patents on life and recovering of the generic basis of life as "commons" which cannot be owned and privatised.

The African countries have also made this demand in W.T.O. Technologies of dependency and corporate control one of the key areas of globalisation has been the entry of global corporations in the seed sector. Control over seed is control over the food system. Besides con- trolling seed through patents and intellectual property rights, corporations like Monsanto and Cargill (now owned by Monsanto) are using technologies of dependency to force farmers into seed dependency. The new seeds are non-renewable they yield a crop, but the crop is useless for seed because of hybrids based on male- sterility, and the impending introduction of "terminator" seeds seeds which are engineered to be sterile. Corporations are breeding seeds that need chemicals life herbicide resistant varieties. Even varieties like Bt. Cotton, need more pesticides because they are prone to more diseases and pests.

The Case of Bt. Cotton

The commercial planting of Bt. Cotton in the country took place, but the cotton crop failed and the government did not clear its cultivation in North Indian states. Leading agricultural scientists, experts and academicians in India have acknowledged the failure of Bt. Cotton on the following counts: drying and falling of squares without boll formation, reduced boll formation, small sized bolls, very short staple length, very little resistance to boll worm, and requiring sprays for control of boll worm, not resistant to dry spells, low yields, low market value and cost-benefit ratio not on par with non-Bt cotton.

A study by the Research Foundation for Science, Technology and Ecology con- ducted in Maharashtra, Madhya Pradesh, Andhra Pradesh and Karnataka showed that not only did Monsanto's cotton not protect the plants from the American Bollworm, but there was a increase of 250–300 per cent in attacks by non-target pests like jassids, aphids, white fly and thrips. In addition, the Bt plants became prey to fun-gal diseases like root rot disease or fusarium. The Bt. cotton varieties gave very low yields. Even the staple lengths of whatever little cotton was produced were so short that it fetched a very low price in the cotton market. Corporate seeds are creating a deep crisis for farmers because of high costs of seed, high dependence on costly inputs (pesticides, herbicides) and high levels of unreliability and crop failure. While costs of cultivation are higher because of costly non-renewable seeds and chemicals, the price of agricultural produce is collapsing due to removal of price and import regulation, a direct result of trade liberalisation.

As production becomes more costly, and agricultural prices fall, indebtedness and economic displacement become the fate of the peasantry, played out in its most tragic expression through farmers' suicides. More than 20,000 farmers have committed suicides because of the debt trap of high cost seeds over the last few years. Farmers are spending Rs. 100,000 per acre and in a good harvest earning Rs. 10,000, a recipe for debt and suicide. Peasant survival, just and remunerative prices, conservation and sustainable use of vital natural resources, including soil, water and biodiversity, healthy food pro- duction, maximisation of nutrition per acre and remunerative farmers' incomes per acre demands that farmers shift from technologies created for corporate control and maximisation of corporate profits to eco-friendly, farmer friendly technologies such as organic farming.

From food first to trade first: How corporations are taking over Indian agriculture through trade Trade liberalisation and globalisation of agriculture is robbing the peasants and landless workers of already low incomes and livelihood security. There are three mechanisms by which the rural producers are either being super-exploited or they are being rendered dispensable.

Destruction of the market support at domestic level both in terms of procurement and in terms of guarantee of Minimum Support Price (MSP). The MSP is to agriculture what minimum wages are in the individual and service sectors, the minimum prices a farmer should receive to cover costs of production and her/his labour. However, as a result of globalisation, the government has started to withdraw from its role in procurement and price regulation. Riots have occurred in different parts of the country with farmers protesting against lack of markets and fall in prices of agricultural commodities.

Diversion from food crops to perishable cash crops and promotion of mono- cultures thus creating market dependency on corporate monopolies. Globalisation policies have promoted the idea that farmers should shift from food grains and stapes to vegetables and fruits. Export promotion zones for fruits and vegetables are a major thrust area in the new agriculture policy. On the one hand this erodes food security for households and the nation. On the other hand it pushes farmers into distress sales, since fruits and vegetables cannot be consumed or stored at the household level.

Removal of Quantitative Restrictions (QRs) on imports and dumping of subsidised, artificially cheap imports. A dispute initiated by the U.S. against India in the WTO forced the removal of QRs. This has translated into destruction of domestic markets and prices in India being perturbed by the artificially low international prices of commodities. While forcing India to remove import restrictions and reduce domestic support to farmers and the poor, the US has further increased its farm subsidies to USD 180 billion over the next six years amounting to USD 20 billion annually. Most of these subsidies go to agribusiness and to capture export markets. In addition, $10 million funding was made available for export promotions by 65 U.S. trade organizations under the 2002 Market Access Programme (MAP) and another $90 million for Market Access under the fiscal year.

The Farm Bill 2002 also provides that MAP funding be increased to 200 million. Transnational agribusiness giants like Cargill, ADM and Conagra are, in fact, the only beneficiaries from the liberalization of imports and removal of import restrictions. They benefit both from using their immense financial clout to depress world prices during procurement and hike it during sales, as well as from the various subsidies that are given to them for both exporting as well as importing, from both exporting and importing countries. A recently released report from the International Agriculture and Trade Policy Institute has shown that in four major U.S. commodities, the level of dumping has increased since 1995 when the W.T.O. came into force, even though the proclaimed aim of W.T.O. is to "reduce distortions in trade".

From 1995–2001 dumping jumped from 23–44 per cent in the case of wheat, 9–29 per cent in the case of soya beans, 11–

33 per cent in the case of maize, from 17–57 per cent in the case of cotton. According to the World Bank, low cotton prices in U.S. resulting from high subsidies are costing African countries $250 million each year. 4.1. The case of edible oils From 70s to the late 80s, India was a heavy importer of edible oils. India produced 3.9 MT of edible oils, and imported 1.5 MT. However, thanks to the Technology Mission on Oilseeds, the total oilseed production soared from 11.3 MT to 21.5 MT in 1993 (10.5 per cent average annual growth).

Imports fell to a negligible 0.35 million tonnes. As a temporary shortage of edible oils, combined with unchecked hoarding drove up prices, the government liberalised imports. At the same time the US flooded the world market with soyabean and soya oil, further driving down international prices of soyameal and all edible oils. As a result India has become the largest edible oil importer in the world. 43 per cent of the total edible oil available in the country is imported. As the acreage under oilseeds decreases, with farmers reeling under the price collapse, further imports are envisaged.

Thus, while removing of subsidies and creation of a "level playing field" was the most important argument used by the government for joining WTO, northern subsidies have actually increased and the playing field has become more uneven. The removal of QRs when combined with dumping becomes a genocidal trade system in which small peasants are wiped out to create global corporate monopolies over food, resulting in the dismantling of the domestic food production system. In the face of rising subsidies and increasing dumping, import restrictions and countervailing duties are a right, a survival necessity, as countries like India, Argentina, Philippines have proposed. WTO has robbed countries of this right through Art.

India and other developing countries should focus on stopping dumping by eliminating Art of Agreement of Agriculture (AoA) which is the basis of the destruction of food security and rural livelihoods in the Third World. Once this crippling clause is removed, countries can start building a global system on citizens initiatives and national priorities that ensures sustainability, supports small farmers, ensures just prices, prevents dumping, protects the countryside and the environment and ensures good, safe, adequate food for all.

Globalisation and the creation of hunger Agribusiness is harvesting both the stocks and agricultural subsidies to gain through agricultural exports. The same companies that gained through deregulated imports are now also gaining through deregulated exports. Millions of Indians go to bed hungry, as the government has dismantled the PDS system, and taken affordable food out of the reach of millions. Foodgrain export is liberalisation's answer to the problem of "over-procurement" from farmers while it will bring in much needed foreign exchange. However, ever since Indian agriculture was subjected to liberalisation, both the share of agriculture exports in India's total exports as well as their value in hard cash has actually declined. The share of agricultural exports in India's total exports declined from 20.33 per cent.

Foodgrain exports did not benefit the national exchequer each period of large exports has been followed by acute food shortage, and India has had to import the same grain again, often at higher prices. For instance, India exported 2.5 lakh tonnes of wheat during later nineties, followed by an import of 18 lakh tonnes in 1988–1989. Foodgrain exports did not benefit the farmer Wheat exports had an inflationary impact on domestic prices, although this did not benefit the farmers, as during the harvesting period the prices are kept low, and even MSP rates are not enforced.

Foodgrain Exports

The push towards exports is neither aimed at helping farmers or the nation's finances; rather, in accordance with the impetus of WTO's Agreement on Agriculture, it is a way of diverting support away from farmers towards traders. While the government dismantles procurement and public food distribution because they are considered subsidies to the people of the country, and are supposed to distort trade, it allows traders to buy bulk grains from its godowns at prices even lower than those offered to the poorest of the poor.

Moreover, the annual budgets since liberalization having been adding to the subsidies for the corporate sector–tax holidays for building silos and cold storages, incentives for exporting, subsidized transportation to the ports of the traders' choice.

Globalisation, as directed by the World Bank/IMF structural adjustment programmes and the WTO has seen a return of famine. The intensification of hunger over the last five years, even while grains are overflowing from the godowns can thus not be attributed to drought and climatic conditions alone, but to the policies that govern people's access to food. The policy changes have consistently attempted to decrease the government's role in ensuring food security and livelihood security for farmers in particular by calling measures meant to aid people as 'trade distorting' and demanding that these be scrapped. The anomaly of the co-existence of burgeoning food stocks and mass starvation is today, as in the late 19th century, the direct consequence of trade-driven agricultural policies.

The 1877 Deccan famine killed over a million people; the trade liberalisation of the last decade of the 1990s has seen India step into the new millennium with over 50 million starving people and millions more with drastically decreasing access to food. An obvious example of the policy shift from people centred concerns to trade- and corporation-centred concerns is the fact that while farmers are not allowed by law to take their produce beyond their state borders, traders can pick them up any- where and take them anywhere. In fact, the government is building super highways, after forcibly taking away land from farmers and communities to connect centres of agricultural production to airports and ports, so that corporations can quickly transport these commodities for export.

The policy changes induced by trade liberalisation include:

- Dismantling the Food Corporation of India (FCI) and reduce its role in procurement from farmers; removal of QRs on imports of food and agricultural products;

- The amendment of the Public Distribution System, to cater only to the BPL category of the population; increase in the central issue price, from Rs. 450/quintal in April 1995, to Rs. 682 quintal in April 1999, and to Rs. 900/quintal in 2000; and
- Increased price of food available through the targeted PDS to the BPL as a result of transferring 50 per cent of the procurement and distribution costs of the government to this category. These changed policies have had disastrous impacts for both farmers and consumers.

Farmers

Government procurement centres refusing to purchase foodgrains from farmers. The refusal of the government to enforce MSP on private traders and corporations, forcing farmers into distress sale of foodgrains at far below production costs. Dumping of cheap, subsidised agricultural products by other countries in the Indian market. Increasing price of farm inputs, including seeds. Liberalising seed regulations to allow private seed companies sell uncertified seed. Deepening of farmers' debt, increased mortgages and land alienation, increased destitution, suicides and sale of body parts among farmers.

The prosperity that globalisation was suppose to spread is fast proving to be elusive. Trade liberalisation and globalisation has resulted in thousands of farmers sacrificing their lives and livelihoods. In fact, the most prosperous state, also called the breadbasket of India, Punjab, has left behind Andhra Pradesh in the notorious distinction of farmers' suicides.

Consumers

Massive reduction in the number of people accessing food from the targeted PDS due to the inability of the government to identify the BPL category of people. Inability of the majority of even the few BPL identified to purchase food-grain from Fair Price Shops due to increased prices. Drastic increase in food prices—by over 60 per cent since the initiation of trade liberalisation measures, and over 200 per cent in commodities like pulses. Reduced PDS off- take by states due to increased issue prices.

The shift from PDS for all to Targeted PDS was justified on grounds of reducing government expenditure. However, with trade liberalisation, the PDS costs to government have risen from Rs. 5,166 crores in the mid 1990s to Rs. 9300 in 1999–2000. 3 While the government blames farmers for this increased expenditure, the primary reason is the increased cost of food to consumers as a result of policy changes, resulting in a drastic decline in purchases from the Fair Price Shops. Declining food production Export-oriented agricultural policies that push the small farmer to destitution on the one hand, and promote cash cropping on the other, have resulted in a steady decline in food production since the early 1990s.

The collapse of domestic support for food production (through dismantling the MSP, rising costs of inputs, crop failure due to uncertified seeds) in the late 1990s has intensified this shift, as farmers are desperate to recover their losses. For instance, there has been a decline in food production to the order of 12.8 per cent in just one year, 2000–2001.

Declining Food Consumption

A major impact of trade liberalisation policies has been a general lowering of food consumption. The per capita cereal consumption has declined from 17 kg per capita per month in the 1950s to 13.5 kg per capita per month in the 1990s. The National Nutrition Monitoring Bureau 1997 data shows a declining trend in consumption in rural India, particularly in cereal and millets, the main source of energy for the poor, from 1990–1995. The National Sample Survey (NSS) rounds starting from the 38th round have documented the decline in cereal consumption 1992, following the implementation of SAPs.

The most important reasons for the decline are: rising food prices, destruction of livelihoods, destruction of the PDS system and shift to export oriented crops. The reduced consumption on the one hand, and decline in production and procurement on the other are directly linked though the food and agricultural policies of the government. Colonialism had destroyed the food sovereignty of the country, forcing changes in cropping from food for local and regional food security to commercial crops. Rice particularly had become a commercial crop even within the country. The emphasis on foreign trade had resulted in large scale famines in the country, forcing Cornelius Walford to comment in 1879 in The Famines of the World—"it is an anomaly that, with her famines on hand, India is able to supply food for other parts of the world".

Following independence, the Government's priority was to ensure that farmers would produce food and thus government procurement (to ensure both that farmers produced food, and got just price for it) and the public distribution system (to ensure that consumers got adequate food at affordable prices) were designed. The need for government involvement in food production and distribution became even more necessary with the Green Revolution that firstly, destroyed regional food security based on diverse cereals and replaced it with just wheat and rice; secondly, concentrated the production of these cereals in just two states—Punjab and Haryana—and thirdly, forced the farmer into the vicious treadmill of costly input seeds, chemicals water intensive agriculture.

Creating a Hunger Free, Suicide Free India

The future of food and agriculture in India appears dismal under the rules of globalisation which promote corporate control. The corporate future will ensure that landlessness will grow, farmers will be transformed from owner-cultivators to serfs controlled by contracts, seed and water will be the private property of global corporations who will sell it at high cost to farmers. In the future based on WTO rules, hunger will grow, and governments will fail to guarantee their people the right to food and their farmers the right to food sovereignty. But this future is not inevitable. Other futures are being created, based on alter- native principles and practices which will over time bring policy shifts.

Principles and policies for the future of food:

1. *Food sovereignty and food security:* Every person, every community, every country has a right to be able to produce to food or have access to food and to shape their food and agriculture economies.
2. *Right to food is a fundamental human right:* Food is the first and foremost source of sustenance, and only a secondary commodity for trade. The Right to food is a fundamental right.
3. *Right to livelihood security for farmers:* The right to work is a fundamental right. For peasants and landless workers, this translates into livelihood security in agriculture.
4. *Right to land:* Land rights are central to food sovereignty. Land reforms interrupted or reversed by globalisation need to be reintroduced and made central to peace and security of the people.
5. *Right to seed/farmers' rights:* Farmers are the first and last breeders. Farmers' rights are collective rights based on collective, cumulative innovation. Farmers have a right to seed and to be free of seed monopolies. Biodiversity must be conserved and protected as a commons. Corporations are liable for seed failure and genetic pollution. The polluter pays' principle must be applied to genetic contamination of crops and food.
6. *Right to water as a public good/commons:* Right to water is a community right and a fundamental right. Farmers and communities have a duty to conserve and share water and manage it as a commons. The polluter pays' principle must be applied to water pollution.
7. *Sustainable production:* Sustainable agriculture is good for the environment, conserves biodiversity and water, and reduces costs of production for farmers. In terms of overall output, it has higher productivity.
8. *Decentralisation:* Decentralisation is becoming an ecological economic and political necessity. Climate change is making centralised resource systems, which

contribute to greenhouse gases, more vulnerable to breakdown. Centralisation in the hands of a few corporations is also responsible for exploitation of farmers and collapse of farm prices. Decentralisation requires priority to localisation of markets and distribution.

9. *Diversity:* Biological diversity of farms and agro-ecosystems and cultural diversity of food systems and cuisines produces more, better and healthier food and richer cultures.
10. *Food safety and food quality:* The industrialised globalised food system is creating food hazards and unhealthy foods. Food safety and food quality requires ecological production, decentralisation and diversity, instead of chemical production, centralisation and monocultures.
11. *Small farms:* Small farms produce more livelihoods and more food than large industrial farms and manage resources more sustainably. The future of agriculture must be based on small farms and small farmers, who are the backbone of agriculture.
12. *Support and subsidies:* Support for agriculture is a food security imperative. Public support must be directed at sustainable production and a fair market, not for corporations and traders, toxic chemicals and GMOs and unhealthy food production.
13. *Just and fair prices:* Agriculture prices are falling below survival levels to sustain small peasants and landless workers. Food prices are rising and taking food beyond the reach of the poor. A floor must be guaranteed to ensure just prices to farmers. A ceiling must be ensured to keep food accessible to the poor.
14. *Just and fair trade:* Current rules of trade benefit corporations and destroy farmers. They promote dumping and deceitful and forced trading. Trade rules must be rewritten to ensure food first, not trade first. Countries have a right to restrict imports, protect livelihoods and ensure food sovereignty.

15. *Food not trade first:* Trade first policies have reduced food to a commodity, reduced agro-biodiversity to 7 globally traded crops with a handful of corporations controlling input and trade. Food and nutrition need to be put first in agricultural policies.

Practices that are Creating Another Future for Food

Amidst the tragedy of farmers' suicides and hunger deaths induced by policies of globalisation, another agriculture is being born in the country, building on the knowledge and wisdom of an agriculture that has survived over millennia and sustained millions for livelihoods and food. Ecological agriculture and organic farming are transforming the negative economy of high input industrial agriculture into a positive economy based on internal inputs. Direct marketing and fair trade are increasing the incomes of farmers and bringing consumers safe, healthy and afford- able food.

A number of initiatives in various parts of the country like Navdanya and its partners across twelve states are reversing the hopelessness of highest seeds with their own seeds of hope. The "Living Democracy" movement is committed to keeping seeds, biodiversity and knowledge of the commons and creating a Bija Swaraj—Seed Sovereignty. The national Jal Yatra and the Jal Sansads are spreading the movement to conserve water and defend community rights. The seeds of Jal Swaraj are being born from the crisis triggered by water destruction and water privatisation. Community seed banks and biodiversity conservation initiatives have saved seeds, and through them the freedom and prosperity of farmers.

The people's movement in Ralegaon Shindi, Maharashtra had single-handedly reversed desertification and economic collapse. Local residents had built water harvesting systems made up of small dams, and they are now growing crops worth $146,000– 188,000 dollars a year. In Gujarat's 13,000 water-starved villages women members of water councils are taking the lead in creating water-harvesting systems. The people's investment in water conservation has also helped recharge ground- water, fill rivers, and increase crop production.

For instance, in 1994, the Arvari River came back to life as result of recharge by 500 johads, the traditional tank system for water harvesting. Similarly, Ruparel, once a dead river, has been flowing since 1994 and is now the leading source of water for 250 villages. It was replenished by 250 johads. In the Alwar district of Rajasthan, the youth organisation Tarun Bharat Sangh mobilised people to rebuild johads for which local communities contributed $2.2 million and built 2,500 tanks in 500 villages.

The collective decision-making process over-construction, maintenance and use of water system has helped prevent conflicts. Tarun Bharat Sangh received in 2001 the Magasasay Award for its work in water conservation. The above trends show that agriculture which protects the health of the soil, health of plants and animals and health of people is emerging. This agricultural revolution is not just happening in India. It is taking place worldwide. Two hundred sustainable agriculture initiatives in 52 countries which have grown over the past decade, covering 9 million farmers and 29 million hectares has led to an average of 93 per cent increase in per hectare food production. This is much higher than what the green revolution achieved, or what genetic engineering claims to offer. Ecological, sustainable agriculture, does not just protect the environment and small farmers livelihood, it also produces more food. A better agriculture is not just possible, it is happening. The future of food is emerging on small, ecologically managed biodiverse farms.

11

War and the World's Food Economy, Agriculture, and the Rural Community

CONSOLIDATION OF FOOD RESOURCES

AFTER seven years of rearmament, the absorption of Austria, and the partition of Czechoslovakia, Hitler felt that he was ready for his big adventure. He invaded Poland in September, 1939, four weeks later in the season than World War I had begun. Then the grain harvest was more certain to be in the bins in Germany as well as in Poland. The motorized blitz against the eastern neighbour was largely designed to strengthen Germany's position. When Austria was annexed in March, 1938, it added to Germany's food balance sheet a deficit of several hundred thousand tons of grain and several hundred thousand hogs a year. The absorption of Czechoslovakia into the Nazi orbit in March, 1939, gave the Reich industrial resources and military assets. Agriculturally it meant merely a deficit of over 100,000 tons of vegetable fats and a surplus of some 300,000 tons of sugar and some lumber.

But Poland's position as the booty of aggression was different. As an agrarian exporting country with a great potential production capacity whose actual output of food had been held back by tight export markets, Poland promised to become a valuable economic asset. Overabundance of man and woman power, a great capacity for producing grain, potatoes, sugar beets, oilseeds, cabbage, hogs, butter, eggs, flax, and hemp, and a vast exploitable stand of timber made conquest very profitable. Several provinces, such as Pomorze and Posnan, had once been German territory, and were familiar ground to thousands of Polish-speaking German farm managers, thus making the seizure of the resources even more feasible. After a brief but heroic stand of 46 days, the most valuable part of Poland fell under German administration — before the first three months of the war were over. Russia annexed the eastern provinces of Polish White Russia and the Polish Ukraine.

Denmark and Norway were overrun. These two countries incorporated important food resources into the nazi war economy. It is wishful thinking to consider both countries, simply because they imported much grain and fats, as an obligation in the German food ledger. Denmark was one of the world's leading quality exporters of butter, eggs, and pork products, shipping the greater part of her exportable produce to England. Danish agriculture therefore brought a welcome windfall of food to the Reich, particularly as long as the large staple of livestock could be reduced. Beyond that initial phase, the well-equipped and competent farmers could adjust their production to less imported feed and to the special demands of the German market. Norway as well as Denmark were considered by the Germans to be an important source of fresh and salt fish. Moreover, with her grip on Norway, Germany controlled the world's greatest exporter of cod-liver oil, still one of the excellent concentrates of the antixerophthalrnic vitamin A, and the antirachitic vitamin D.

The incorporation of all Holland, Belgium, Luxemburg, and the industrially most important part of northern France into the Reich's orbit. Belgium, Luxemburg and France were industrial and strategic gains. As for food balances, Belgium at least added a considerable deficit in grain, fats, dairy products, and meat. Holland, on the other hand, meant a definite gain in food resources, as had Denmark.

When Italy joined Germany as a belligerent ally, she did not assist in food supplies, but promised to ship citrus and other fruits as well as some winter vegetables to Germany in exchange for coal. The basin of the lower Danube was incorporated into the *Grossraum* or Continental block economy, partly by treaties, as in the case of Hungary, Bulgaria, and Rumania, partly by military conquest, as in the case of Yugoslavia. These countries comprise a group that exports food, particularly cereals and meat, but also exports a sizable volume of oilseeds, fresh and dried fruits, vegetables, and eggs. This group also exports much tobacco. Greece, which after conquest was also merged into the Continental block, is a wheat, meat, and sugar importer, but normally exports olives and olive oil, raisins, figs, dried currants, and tobacco.

The four Danubian countries combined had for the average an annual net export of 3.6 million tons of cereals, 400,000 tons of animal products, 175,000 tons of oilseeds, and 114,000 tons of fruits. Greece imported for the same period 584,000 tons net of wheat and 330,000 tons net of meat, but also exported net 29,000 tons of oil and 128,000 tons of fruits. The four Danubian countries plus Greece therefore added to the Nazi war food economy a normal net supply of 3,000,000 tons of cereals, 242,000 tons of fruits, 204,000 tons of oilseeds or their equivalent in oil, and 70,000 tons of animal products.

The difficulties of intra-Continental transportation require some deduction from such exportable surpluses of prewar years in an estimation of the value of conquered countries for the fortress economy, particularly as the war's strain bears down more on the railroads. Yet, it would be a mistake to exaggerate the deductions on that account.

Only 20 months after the invasion of Poland, the European Axis controlled practically the entire Continent up to the new boundaries of the Soviet Union which, after the annexation of Estonia, Latvia, Lithuania, eastern Poland, and Bessarabia, ran almost in a straight line from Memel to the western part of the Black Sea. The Axis territory included Norway and Finland. It was contiguous to Spain and to Turkey, both of which traded, as they still do in 1944, some food and many important industrial raw materials with Germany. Coastal shipping all around the Continent was under Axis control except for the coasts around the Iberian Peninsula which lay under the shadow of Gibraltar. From the summer of 1941 to November, 1943, Axis territory was expanded far to the east. After the attack upon Soviet Russia the eastern boundary was rolled back to Leningrad and partly beyond, close to Moscow, and from there south, swinging much farther east to the Volga and deep into the Kuban area, almost to the Caspian Sea.

This steam-roller campaign of mechanized armies over the vast expanse of Russian plains yielded the three Baltic states, exporters of livestock products; eastern Poland, a grain, potato, flax, and pork producing area; Bessarabia and the Ukraine, with a normal surplus of grain, sugar, oilseeds; and finally the lower Volga and northern Caucasus districts' grain, oilseeds, and meat.

From the beginning of the war, the German war food administration took it for granted that the European Continent would be cut off from other continents by blockade, and that the food requirements could be covered only by drawing upon Continental resources. Occasionally a few blockade runners came through with food. A daring surface raider in 1940 brought from the Antarctic a whole Norwegian whaling expedition, the floating factory's tanks bulging with whale oil. Such dribbles in their sum total, however, never reached a significant figure.

THE FORTRESS FOOD ECONOMY

The Nazi administration has managed the food economy of the Continental fortress by devices that are a part of the ordinary machinery of a planned and regimented economy. Large stock piles of all essential commodities accumulated before the war have been carried over in order to meet temporary emergencies and shortages. Production in Germany as well as in all the occupied territory is directed according to national production goals, is stimulated by generally profitable prices, its progress is constantly surveyed down to the last farm. The goals are fixed annually, and determine whatever changes in acreage of specific crops are to be made and how large the herds of livestock and the output of animal products shall be.

The delivery of all products from the farm to the proper agencies in the market is subject to public control as to time, volume, and destination of delivery. On the demand side of the market a similar system of public planning and controls operates. After ample reserves are set aside, the available annual output of all foodstuffs is allocated to major groups of claimants according to necessities of warfare and priority rank. While the armed forces claim their allotment of food on the basis of established army diets, the requirements of all civilians are tailored to fit available residual supplies. Civilian consumption is either strictly rationed, or, where this is not feasible, consumption is restricted by curtailment of available supplies. The same principle of public control of demand applies to the utilization of food or feed products and to fertilizer as well.

In other words, not only the people, but all the animals and even plants are put on rations. Any sort of utilization of agricultural produce is subject to central controls. In principle this system is airtight. It almost eliminates the element of surprise and uncertainty in consumption, demand, and utilization, while it still must cope with the fluctuation of yields and crops due to weather and with many other biological factors beyond human influence.

In comparison with the strategy used in the American war food administration, the German administration has made maximum use of the price incentive by fixing such differentials that production goals are enforced.

All the available evidence convinces the author that the thesis he held and made public from the outset of the war is still valid, and probably will remain so: the Axis cannot be defeated by food shortage, but must be defeated on the field of battle.

This considered opinion, formed on the basis of years of economic analysis and observation, in no way implies that populations everywhere in Europe or even a majority of the people could maintain anything like their habitual peacetime diet, or that agricultural production could be protected against the effects of total war and could continue to produce as in prewar years. What has happened in more than four years of war behind the blockade cannot be condensed into any simple formula. It requires the careful consideration of the many multiple facets of the food situation. While a great amount of information is at hand on many details, nowhere has an attempt been made to compose a more general picture of the European food economy during the war, sufficiently concise and accurate yet free from detail that would swell it to encyclopedic form. Following is an attempt to fill this gap at least tentatively. It is confined by the requirements of our subject, the reconstruction of world agriculture.

As soon as Germany had accomplished the first part of her bid for Continental empire by military conquest, she began to consolidate the accumulated resources, to reorganize their utilization, and to mold them into an integrated system of planning and regimentation.

Indeed, the nazis did not wait for the completion of conquest, but began to set the wheels of reorganization in motion a few days or weeks after the armies had swept through. It was reliably reported that three days after the Germans had begun to occupy the Sudeten area, German soldiers in uniform were seen plowing the fields. In Poland, the farm-management personnel of the army's *WirtschaftsStabe* followed immediately behind the combat forces and began to direct farming operations. There is much evidence that the same pressure to re-establish farming to full activity was exerted in all occupied areas. The methods used differed from country to country. Moreover, each country was assigned a particular place and task within the food economy of the Continental fortress.

The Continental Food Economy

There is no scheme of the Nazi blueprint at hand, but the major outlines of the plan for the organization and the performance of European agriculture can be drawn. A welter of articles in European newspapers and magazines especially from neutral countries, reports by high Nazi officials, and orders and ordinances provide the evidence. Since they had to face the possibility that the conquest might be disputed at any time, either by invasion and air bombardment from without, or by mass revolt and sabotage from within, the nazis conceived their Continental food economy as a system of concentric rings, corresponding to successive rings of military lines of defense.

Agriculturally these rings proceed from the outermost zones, which may or may not yield some supplies beyond those needed solely for their populations, to zones of increasing importance for the import requirements of the Reich, with the Reich itself forming the centre. Naturally these areas are not evaluated exclusively as sources of agricultural products. Their economic rank is decided by the aggregate of all materials and services to be derived from them, as well as by their military importance. Considerations of food requirements for the conquered people are subordinated to those of securing the essential war supplies for the nazi machine.

Thus the inner ring of the fortress consists of the Reich of 1937, Denmark, Holland, Belgium, the French *départements* east of the Maginot Line, including Alsace-Lorraine, Austria, Czechoslovakia, Poland except for Polish western Ukraine and Polish western White Russia, Lithuania, Latvia, and Estonia.

This central pivot embraces a great range of agricultural conditions. It has a decided slope from west to east in productivity of soil and the skill of men on farms, but it has also a concurrent slope from densely settled and highly industrialized areas in the northwestern corner and the west in general, toward more rural and sparsely populated regions in the east. Therefore the eastern belt from Estonia down to Moravia counts heavily as a food surplus producing area, especially for animal products, although the Corridor and Posnan can also export sizable quantities of potatoes and rye. Austria's and Czechoslovakia's industrial resources, located

— at least for several years — well out of heavy bomber range, are of such importance to Germany's war potential that they are considered vital parts of the Reich. If they need food, they will get it, just as any German province would; and if their normally good capacity to produce food can be expanded, they will receive the means with which to expand it.

Belgium and the adjacent corner of France are in the same class. The Brie-Longwy coal basin and the heavy industries in it are an extension of the Ruhr and the Saar districts. For that reason their food deficits will be covered by the German food administration. At the same time, every effort is being made to step up and adjust to war needs the local production of food, particularly since the climate and soil are good, farmers very efficient, and local produce can find its market without involving a great deal of transportation.

Denmark and Holland are the two most productive food provinces Germany has. They rank equally high or higher with the lower Rhineland because of their unequaled efficiency and their professional versatility in agricultural production. It is very deceptive to measure the agricultural plant of these two nations simply on the basis of cultivated acreage, livestock numbers, or similar statistical data; and it is similarly misleading to focus mainly on the peacetime dependence of these countries on overseas imports of grains and oilseeds. The German General Staff's *Wehrwirtschaft* section knows better. It is aware that the average Dutch or Danish cow yields in marketable dairy products at least twice as much as the average cow in Poland, and that therefore the 6,000,000 Dutch and Danish cattle yield for the market far more milk of an incomparably better grade than Poland's 10,000,000 cattle.

The war food economy in this central ring, marked "I" on the map, is so managed that it is much better controlled than all the other rings. In greater Germany, i. e., the old Reich plus AlsaceLorraine, Austria, Sudetenland, and the annexed northern part of Poland, the so-called *"Warthegau,"* Danzig, and Memelland, agriculture is incorporated in the *Reichsnahrstand* and its production and marketing are ruled by commodity cartels. The system of food planning, made foolproof before the war, was simply extended to the annexed areas, and experienced administrators transferred to them. It soon began to perform as well as it had earlier although in many areas it did not succeed in keeping production up to prewar levels.

Denmark, Holland, and Belgium were left more or less to self government in agricultural affairs with advice or instructions issued only from the top. In Denmark and Holland the most competent administrations, the farmers' co-operatives, were not only in a position to perform any agricultural planning job much more skillfully than any German occupation authority, but they were keenly interested in doing so, because co-operation promised to be the far lesser evil. Belgium, on the other hand, was a net importer and was consequently less experienced in agricultural planning, so a new war food administration had to be organized.

In the area east of the Maginot Line running from the Channel to the Swiss border, a different system of managing the food economy was established. It is essentially the one applied in the east. The German administration, in view of the absence of any workable part of a national agricultural or food administration for the thousands of abandoned farms, whose occupants either fled or were evacuated, and faced with the Frenchman's suspicion and animosity toward governmental interference in any way whatsoever, put all the vacant land under the management of its eastern farm management corporation, the Ostland G. m. b. H., a corporation under the jurisdiction of the *Reichsnahrstand* in autumn.

This organization put 2.6 million hectares of land under cultivation, using French prisoners of war for labour and large-scale mechanized farming methods. Merging of fields, plowing down old property lines and consolidating the land for tractors and steam plows was to the German managers simply the difference between allowing land to lie idle or using it for producing food. What it did to the French villages and the people who had lived on the good inherited earth was of no concern to the nazis. The German food administration was satisfied if even between the forts and on top of the Maginot Line food was grown once more.

In the eastern part of the central ring, agriculture had to be similarly managed, but it did not present insurmountable obstacles there, either. The "Corridor" and the province of Posen, which had been German up to 1919, were annexed and integrated into the German agricultural fabric. The rest of Poland was organized as the territory under the "Government General" seated in Cracow. All farms over 5 hectares, 3,900 of them, were taken over by a farm-management office which established a centralized direction and control. All farms under 50 hectares, the Polish peasant farms, were under the authority of chambers of agriculture, with a German commissioner in charge, but otherwise manned by Polish personnel. The marketing of all products was organized on the German cartel system. A central agricultural office became responsible for supplying all the farmers' needs and the delivery and sale of all farm produce. The existing system of agricultural co-operative associations was much expanded. In the Baltic states a similar system was applied, leaving the production in the hands of the family farmers, but controlling their supplies and sales, as well as planning production from central offices under German management.

One major gap in the central ring remained. Many large-scale farms were left without owners, occupants, or managers. They comprised either large estates, whose owners had been killed, interned, or had fled, or groups of farmer-peasant holdings consolidated by the Soviet government as collective farms. Most of these large-scale farm units were turned over to the Ostland Corporation, mentioned in connection with the French belt east of the Maginot Line. This corporation operated no less than 5,000,000 hectares of tillable land.

The Outer Rings

The second ring of food resource territory is made up of two areas which compose the greatest contrasts found on the Continent: first, the four Danubian countries, Bulgaria,

Hungary, Rumania, and Yugoslavia; second, Norway. Norway is an excellent source of sea fish, particularly of salt herring, of cod-liver oil in vast bulk, of sizable exports of nitrogenous fertilizer, of exports of aluminum, of many essential minerals and ores, and of wood pulp. The Danubian countries occupy so important a place in the nazi food plans because they are traditionally Europe's greatest exporters of grain, and potentially a fruitful source of oilseeds. Danubian exports of animal products, fruits, and tobacco, are likewise significant.

Norway's fisheries operate under German control on boats and ashore, but Norwegians, and not Germans, manage and run the business. In the Danubian block, German administration finds itself thwarted at every turn. Hungary and Bulgaria manage to retain a considerable degree of independence, so that whatever food supplies are shipped to Germany leave those two countries as exports on clearing account and in exchange for industrial goods. Rumania and Yugoslavia, torn by domestic strife, suffering from German military occupation, are much more subject to overt German pressure and interference. No matter how disappointing the supplies derived from the lower Danube may have been to Germany, that area is still on a par in importance with the rim of the first ring.

The third ring contains the formerly German-occupied northern part of France, parts of Soviet Russia east of the Baltic, the Polish partition line of Lord Curzon and the territory east of Bessarabia. This ring is an area which may not even feed its occupying forces, but still may yield certain moderate supplies. The population of Ring III does by no means rank as high as that of Rings II or I. It may be allowed very low rations for the sake of draining off exports from it. The actual as well as the potential food production capacity of the German occupied Soviet area has been more of an enigma to the German administration than any other agricultural problem on the Continent.

Northern France, including Paris, is of essentially military importance as the base of air-force activities against England,

and as an anti invasion base for ground forces. It also harbors some important armaments industries. Agriculturally, this area is a nuisance despite the fact that it is the part of France which has a high food output per capita or normally yields surpluses. It does not pay to organize it effectively, although some food is taken for the German forces. Occasionally some food imports are needed. The worst black market in all Europe operates here.

The agricultural aspects of occupied Russia are just as perplexing to the Nazis as the Soviet Union has been as a military power. The Ukraine, the proverbial bread basket, with its rich prairie soil, has had its magic lure for the conqueror. When the south Russian Front was rolled back deep into the Kuban district in 1942, the Nazi press began to report the alleged wastefulness, inefficiency, and low yields of Russian collective farming. There were hints about the great opportunity to increase yields under German farm reorganization, but a cautious tone prevailed as to the time required to improve conditions to the point where worthwhile export surpluses could be created. There were grandiose speculations, however, about the restoration of the Ukraine to its position as the supply centre of a major part of the Continent's grain requirements — all, of course, under the tacit assumption that Germany would be able to keep her hands on the Ukraine indefinitely.

The administrative devices applied to the eastern part of Ring III were modeled after the German agricultural cartel and blended with the experiences gained from the economic administration of occupied territories in World War I. Two high commissioners were assigned supreme authority over the area between the Baltic and the Black seas, one being responsible for the Ukraine, the other for the Ostland — all the eastern occupied territory except the Ukraine, the area of the "Government General," the annexed provinces, and the area between the Bug and the Dniester rivers. The latter territory was designated *Transnistria* and subordinated to Rumania's administration.

In January, 1942, a general ordinance concerning the organization of land tenure in all the occupied eastern territories abolished the "bolshevik" institutions of collective farming and replaced them by laws which granted certain individual property rights to land, and stressed the return to private initiative. In reality no such experiment was contemplated or tried. *De facto,* the collective farms were kept as such, while *de lege* they were "transformed" into farming cooperative associations. The high commissioners divided their areas into general districts corresponding to the areas formerly serviced by the Soviet MTS stations, and reopened these machine tractor and repair stations under an over-all *roof* organization which became the seat of the commissioner. The MTS stations are responsible also for supplying the kolkhoz as well as the noncollectivized individual farmers with hand tools and farming implements. All the necessary materials are sold to the MTS stations through the central selling agency of a cartel of agricultural machinery and tool factories. In the Ukraine, for example, 28 factories producing farm machinery, located west of the Dnieper were reopened and organized as a cartel under the *Landmaschinen-Industrie-Ukraine, G. m. b. H.,* at Kiev.

With that portion of the farm population which remained, agricultural production on former kolkhoz farms and individual peasant farms was thus re-established under the command of "leaders" or managers imported by the German administration. Many Dutch farmers were persuaded if not coerced in various ways to go to the Ukraine as managers of kolkhoz farms or groups of kolkhozi. A respectable number of previously state-owned and operated large estates, the sovkhozi, numerous abandoned kolkhozi and also many abandoned individual peasant farms had to be dealt with. All, or at least as many as could be handled, were turned over to the Ostland corporation to manage and operate. The Ostland corporation had to furnish managers and whatever labour it could find for hire, recruit from the occupied population, military prison camps, or concentration camps.

To grasp the size of the administrative job of the two high commissioners and their respective staffs, it may be said that according to German statements, the occupied Russian territories by the autumn of 1942 consisted of 102,000,000 hectares, 34,000,000 of which previously had been cultivated. This area held a population of 52,000,000 in 1939.

When the Russian armies retreated, they applied the scorched earth policy quite thoroughly, destroying or removing as much of the power equipment on farms as they could, evacuating the most skilled of the MTS station and kolkhoz personnel. Yet the Germans were able to restore farm operations to some extent. It will be seen later that in spite of the promise of this eastern portion of what we have called Ring III in the Continental fortress economy, it was a mirage which vanished abruptly when the 1943 Russian offensive liberated most of the area.

To sum up the organization and administration of the Continental food realm under the so-called "New Order," it is apparent that two main ideas have oriented the blueprints. First, a broadened, consolidated central block holds the vital industries and a reliable more or less self-supporting food economy, and second, this block is supplemented by contiguous food surplus areas between the eastern Baltic, the Adriatic, and the Black seas. The addition of Europe's best ocean shores for deep-sea fishing on the north side of the Scandinavian peninsula is of a more peripheral nature. Condensed into a brief formula, the plan comes fairly close to the concept of a consolidated central European economic block — Friedrich Naumann *Mittel Europa,* so popular in Germany during World War I. It is also in essence the line of foreign policy suggested by Hitler and Von Ribbentrop to the Prime Minister of Great Britain during the years of appeasement — "a free hand in eastern and southeastern Europe."

Ring IV will complete the picture of the distribution of weight, importance, and dependence within the Continental food economy of the nazi regime. It is composed of what was the originally unoccupied area of Vichy France, Italy, Albania,

Greece, the four large islands of the Mediterranean, and Finland. This ring, from the point of view of the conqueror's food planning board, is a marginal zone of little or no importance. It may yield some food exports, or it may just as well demand some imports. Its significance is primarily a military one. Southern France, Italy, Albania, Greece, and the four islands are ideal terrain for seaborne invasion, which by the end of 1943 had cracked the defenses of Corsica, Sardinia, Sicily, and the southern part of the Italian peninsula. In 1944, the northern coast of France lent itself to similar operations. Finland is the northern anchorage of the eastern front. Southern France has yielded some vegetables and fish for the northern cities and the German occupation forces. Italy has shipped some citrus and deciduous fruit and some vegetables to Germany. Greece has shipped fruits and tobacco. Finland has exported principally wood pulp and timber. Italy has received sugar and loans of grain and potatoes from Germany; Finland, substantial amounts of grain. Albania and Greece, which were under Italian agricultural administration, did not receive the necessary volume of imports, although some grain was shipped from German-occupied territories.

It may seem paradoxical to suggest that the nazis classify these predominantly rural areas in Ring IV as an immaterial marginal zone agriculturally speaking, while they look upon industrial Belgium and Czechoslovakia as parts of the most vital Ring I. Yet it conforms logically to the concept of German *Wehrwirtschaft* and *Grossraumwirtschaft*. The best evidence for the accuracy of this interpretation is the absence of any efforts on the part of the nazis even to influence the agricultural or food administration in Ring IV. Only since the surrender of the armed forces and the occupation of northern Italy by the German army, has Germany been forced to supervise the Italian food economy.

Prices, the Vital Instrument of Food Strategy

Next we must learn by what methods and policies the nazis have managed the agricultural resources of the Continent, and what effect such policies have had on the productive plant and its performance. We will need to know something about the probable volume and nature of food reserves, the condition in the food market in various countries, and the degree of under consumption to be expected by the end of the war.

When the advance of the German army brought one country after another and all its agricultural resources into the nazi-controlled food orbit, a policy for obtaining the most productive work and the most compliance with production planning had to be devised. Since the nihilistic nazi regime did not recognize moral principles but avowedly made success the only criterion for the choice of means, the whole range of methods of influencing people, from persuasion and economic inducement to outright terrorization, has been considered legitimate. Nearly every device of persuasion has been applied somewhere in some degree.

However, a survey of the nazi policy as it applied to treatment of farmers at home and in occupied countries reveals the surprising fact that contrary to their cynical philosophy about the human beast and their utter contempt for the profit motive, the nazis have, in their efforts to stimulate agricultural production, relied at least partly on the powerful tool of profitable prices. Knowing only too well what happened in Russia when the peasants refused to comply with the wholesale expropriation of their produce, and when mass collectivization was ordered, nowhere have the nazis relied exclusively on coercion. In Germany, very few farmers were ever sent to concentration camps, and the few who went were quickly released because they were needed on the farm. In Poland and occupied Russia, coercion has been used to a much greater extent than in other countries for several reasons.

Nowhere else had warfare disorganized and disrupted agriculture so much as in these eastern areas. Nowhere else were so many people on collective farms in the position of poor peasants, farm labourers, or their equivalent. Yet even in the east high prices for farm products have been the device most often employed by Germany to get the desired results usually in combination with production and delivery quotas. The official German journal of statistics published in December, 1942, an article that showed how nazi economic planning has changed the slope of grain prices on the Continent. The vicious inflation that has swept over the lower Danube basin has not invalidated the price relationship.

As Chart VI indicates, the slope of rye prices which before the war descended from Germany and Czechoslovakia toward the east and southeast was partly reversed in 1942 and in general changed so much that the eastern regions received much better prices than those prevailing in Germany. A serious setback to this policy has been that war conditions have made many of the goods farmers buy so scarce that good money prices may mean little real purchasing power. Even in the absence of goods, however, good farm prices may mean a great deal to farmers who have to face heavy taxes, interest payments, and high operating expenses. They dread becoming insolvent and losing their farms. Naturally, the price incentive operates best where farmers think in commercial terms and are skillful managers, while it is much less effective among subsistence farmers.

In Germany, agriculture has been kept on a profitable basis during the war, as it was in the later prewar years, aided by subsidies to hold down the cost-of-living index. A similar policy seems to apply to Denmark, Holland, and Belgium. Improved prices paid to farmers in these as well as other countries does not imply that the German consumer has to pay more than before, because the Reich's clearing system for all foreign trade as well as the Reichsbank's power to adjust the exchange rate for the currencies of all occupied countries offers ample opportunity to eliminate such undesirable effects upon the domestic German price level. Domestically the method of subsidizing food has been used in order to prevent a rise in the cost-of-living index.

While prices have thus been the chief means of persuading Europe's farmers to produce as much as they can, it has been only one device among many. Production plans have been translated into quotas of production goals for every farm wherever the agricultural administration was able to reach the individual farmers. In many instances permission to purchase means of production such as seed, fertilizer, motor fuel, or spare parts is granted on the condition that production quotas be met. For some products such as oilseeds a bonus for expanded production is paid in the form of an extra ration of vegetable oil, and a generous allowance of oil cake for feed. In former Russian areas, the promise to return the collective land to individual farmers as private property was made with the proviso that a certain minimum acreage be cultivated.

Since a fair rate of agricultural production in no way promises, much less guarantees, a satisfactory supply of food for the armed forces and the nonfarming part of the civilian population, much more ingenuity was required to institute controls for the delivery of produce into the legal channels of the market. An almost airtight system is applied within Ring I. It requires each farmer to report the output of his products, provides for a check and audit of his report with sample tests of its veracity, and for a schedule of the stipulated quota covering the entire crop year.

These quotas are reinforced by orders that control the way in which food products may be used on the farm. If all the wheat produced minus seed requirements is needed for bread, the full delivery quotas are supported by an order prohibiting the use of wheat as feed enforced by a severe penal code. To make full delivery doubly sure, the price of wheat is fixed in such relation to eggs, milk and other animal products that it becomes profitable to sell all of it, and feeding wheat to animals means losing money. A complete regimented economy can exert pressure in connection with almost every single action of the individual or wherever the producer receives from the government permission to buy rationed goods or services.

Fulfillment of delivery quotas has been made a prerequisite for innumerable permits and allocations of rationed or regulated materials. On the other hand, the farmer had been relieved of the tools which might tempt him to break the law. A maximum of milk is earmarked for the creameries for separation and butter-making. To make certain that farmers will not hold back the milk, separate it on the farm, make butter and sell it in the black markets, the government has expropriated all farm separator machines and has made it a criminal offense to possess one.

Owing to these intertwined premiums, penalties, and multiple pressures, the farmer cannot easily hold back much of his produce to sell in the black market. A vigilant policing and detective system, brutal penalties for all parties to a violation, relatively satisfactory income conditions and satisfactory rations actually on hand for farmers are other safeguards against the disappearance of large quantities of food into the black market.

The forceful control and adjustment of consumption and the maintenance of well-filled public granaries are essential parts of the administrative machinery assuring a smooth performance of Nazi war food economy, just as much as they have assured a steady flow of food in Great Britain and the United States. The methodology of rationing was developed to a remarkable degree of perfection in Germany while preparations were being made for war. Political tyranny permitted the design of a rationing policy that can more or less ignore consumer preference and follow closely the principles of greatest economy and efficient use of a given amount of supplies.

Again it is to Ring I of the fortress that rationing is applied with this rigidity, and even there it is not uniformly applied. The German rationing system and the guaranteed supply of the rations pertains only to the "Greater Reich": the Reich of the Versailles Treaty plus Austria, Sudetenland, the Saar, the annexed Polish provinces, Danzig, and Memelland. In the "Government General" rationing is decreed by the German administration on a discriminatory basis for the sole purpose of squeezing as much food as possible out of that area for the Reich.

In all the other countries of Ring I, rationing is controlled by the national administration. There is no attempt to equalize rations between these occupied countries. Hence the rations and to a greater extent the availability of rationed foods differ widely. And as time goes by the differences alter. Denmark and Holland, for instance, have enjoyed a food situation that for several years was better than that in the "Greater Reich." Bohemia and Moravia fared as well as the Reich. The urban populations of Belgium and Norway had to endure the greatest hardships. The outer rings, except for the eastern part of Ring III, are left entirely to themselves so far as the distribution of available supplies among urban populations is concerned.

What, then, actually happened to agricultural production inside the nazi-controlled *Festung Europa* during the first four years and four months of war? If popular opinion in the United States and Great Britain were correct, European agriculture would have been in a most deplorable condition, its productive plant so far depleted, its output so shriveled that its restoration would demand truly gigantic efforts on the part of the United Nations or decades of gradual recuperation. This public opinion is formulated on the basis of a piecemeal collection of thousands of detailed news items reported in the press and in magazine articles, from which a misleading composite picture is unconsciously drawn. It is bound to be worthless, however satisfying to look at, because there is neither a gauge for the validity of the reported observations or the representative nature of the sample nor any guidance for measuring their weight or real importance.

Fortunately for everyone concerned, especially the victorious nations when war ends, we have a great body of evidence indicating that the actual agricultural situation on the Continent behind the blockade is different from what uninformed conjecture suggests. While it is far from the prewar norm, and far from a generally or wholly satisfactory state measured by modest wartime desires of all consumers, nevertheless Europe's agriculture is far from disastrous decay. The most comprehensive effort to estimate the output of food in Continental Europe during the war is found in the calculations made by John H. Richter, the expert on Continental agricultural production in the Office of Foreign Agricultural Relations in the United States Department of Agriculture. For years the Department has been the world's most resourceful and best informed institution with reference to international agricultural conditions.

According to this estimate, and ignoring the 2.4 per cent of all calories converted for imported food, total Continental production of food in 1941-42 was 4.7 per cent lower than in 1936. The curtailment lay exclusively in the field of animal husbandry and the feed supplies for it. Measured in calories and leaving unconsidered the production based on imported feeds, the output of milk and dairy products decline 22 per cent, that of all meats including poultry 21 per cent, that of hog fats and tallow 19 per cent, and that of eggs 50 per cent. If the production based on imported feeds is included in the output of animal products, the decline is naturally greater. The corresponding data are: milk and dairy products down 29 per cent; meat and poultry down 30 per cent; hog fats and tallow down 31 per cent; eggs down 52 per cent.

For a critical evaluation of this total Continental balance sheet, some basic observations derived from continual examination of current information about individual countries for the years in question may be helpful. Mr. Richter's balance sheet compares the crop of 1941 with the crop of 1936. Figures are available for three Continental crops since then. One cannot operate with simplified trend lines, and conclude from the rate of decline in production recorded for the interval between prewar and the level of the 1941-42 crop year that the same rate of decline has persisted throughout 1942-43, and still holds for the 1943-44 crop year ending on June 30, 1944. Even agricultural economists often overlook the influence of weather, in particular the amount of effective soil moisture. This factor is so powerful that it can overshadow, overcompensate or exaggerate the effect of the production factors under human control, such as application of various fertilizers, methods of cultivation, and even considerable changes in acreage. Lack of comprehension of the wide range of crop yields per acre possible because of the influence of weather alone is responsible for the majority of false prognostications of European war food supplies.

The years 1938 and 1939 were favoured by good weather and saw two exceptionally good crops. They were followed by three years of poor weather in the growing season, and with two of the hardest winters on record. A German agricultural experiment station in the *Nordmark* recorded that frost penetrated 11 centimeters in the ground, 92 centimeters as much as 115 centimeters. In the two previous winters there had been 58 days of frost, but no less than 82. In the latter case all the fall-sown oilseeds, one-third the fall-sown grain, and in many parts of the country almost every fruit tree, were killed by frost. Moreover, spring sowing was delayed three weeks. These hard winters swept over all of northern Europe. The grain crop was damaged also by too much wet weather during the harvesting and fall sowing seasons.

An early spring and warm weather followed a relatively mild winter, but in the critical period of May and early June, when easterly winds prevail, serious drought developed all over the Continent, deep into Russia. For some time it looked as if the failure of the grain crop was certain. The nazi food administration girded itself for the worst, curtailed rations, restricting livestock feeding still further. Miraculously, warm rains began to fall at the last minute. The crops recuperated and continued to improve.

The following tabulation shows the fluctuation of the European wheat crop in Continental Europe excluding Russia, as estimated by the Food Research Institute, in million bushels:

Year	Total	Four neutrals	Others excluding Danube	Lower Danube
1938	1,778	149	1,163	466
1939	1,621	162	1,008	451
1940	1,225	111	819	295
1941	1,345	139	876	330
1942	1,240	152	823	265
1943	1,425	130	920	375

Other crops had better results than wheat in the critical three years. In 1940 the Continent harvested a large crop of potatoes, in 1941 a fairly good one, in 1942 a bumper crop but in 1943 a miserable one, which yielded 20 per cent less than in average years. The important rye crop was considerably below average in 1940; average or slightly above in 1941.

In wartime adjustments the greatest emphasis of crop production has been laid upon an increase in the acreages of oilseeds, sugar beets, potatoes, and truck crops. Oilseeds are sorely needed to relieve the shortage of edible and soap fats because overseas imports have ceased. Increasing oilseed acreage has been especially successful in France, Belgium, Holland, Denmark, and Germany. The Continental acreage of sugar beets, the most labour-intensive and high-yielding crop, was in 1943 expanded by 11 per cent beyond prewar levels (excluding Finland, Sweden, Slovakia and France).

Hungary, Switzerland, Turkey, and Rumania cultivated from 30 to 40 per cent more acres than in 1937-41. Potatoes are the full equivalent of bread grain as well as feed grain, yielding about twice the calories per acre that the grains do. Germany, Denmark, Holland, Belgium, Switzerland, Sweden, Italy, Hungary, and Spain all had in 1942 a considerably larger potato acre-age than in 1939. Truck crops yield at least as highly in tons as potatoes, and contribute to the diet many valuable mineral-bearing foods. Potato culture has expanded almost everywhere, although for the Continent as a whole the increase is not sensational. The most striking increase appears in the expansion of vegetable acreage. All countries where production techniques are developed have made much headway in supplementing diets with vegetables unrationed because they are perishable products. Some German data may illustrate the campaign to expand commercial vegetable production:

Year	Acreage (1,000 ha.)	Crops (million metric tons)	Crops in Per Cent of 1939
1939	138	2.54	100
1940	164	3.08	121
1941	211	4.11	162

1942	313	6.43	253
1943	393	7.10	279

The crop of commercial market vegetables has thus been almost tripled during the war years. Germany, Holland, Denmark, Belgium, France, and Switzerland have made great efforts to dehydrate and can much of the crop at the height of the season to carry it into the winter months. Under the pressure of war even refrigeration has made great strides in western Europe.

Another important expansion in food production, for which no statistical data are available, has taken place in uncounted millions of "victory" gardens. Necessity has persuaded Europeans to a far greater extent than it has Americans to plant vegetable gardens. The contribution to the kitchen from this most intensive cultivation by consumers can scarcely be overestimated. In the United States 12,000,000 more vegetable gardens than in peacetime have produced in 1943 an estimated crop of 4.5 million tons of vegetables.

The Reduction of Livestock Herds

The chief adjustment within the Continental food economy has not, however, taken place in primary production, that is, production of plants, but in the conversion of feed into animal products. This is the most logical and seasonable adaptation whenever the grand total of available food calories is short or threatens to become short. The detour of converting edible vegetable foods into animal products involves a net loss in caloric energy which ranges all the way from 40 to 90 per cent according to the composition of the feed, the type and age of the animals, and the type of animal product obtained. For practical purposes it is advisable to calculate with losses near the upper limit. Such a loss is unavoidable wherever inedible materials such as grass, clover, alfalfa, or fodder beets are fed, but it is plain folly to feed grain to animals when there is not enough bread grain for all the people.

The various nations have decided to reduce some of their livestock herds, thereby making a considerable and actual net gain in available food energy which ranges from 75 per cent in feeding pigs to 90 per cent in feeding poultry. Their diet has thereby become more frugal, and contains less of the delectable items which a highly advanced food economy enjoying international exchange had presented. This nevertheless does not imply that a predominantly vegetarian diet of this sort must necessarily be inadequate, as British public health reports for the war years bear out.

The reduction in livestock so far has remained within the bounds of desirable economic adjustment, and has by no means dangerously depleted stocks. Since the chief purpose of reducing livestock feeding is to save grain for direct human consumption, the numbers of chiefly grain-converting animals must be reduced most. Hence chicken flocks were much reduced, especially where they were customarily fed on grain imported from overseas. By July, 1942, the Danish poultry flock was down to 11,000,000 birds compared with 29,000,000 in 1939.

The Dutch flock was cut much more — from 33 million in 1939 to 3.8 million in 1942, but increased to 3.9 million in 1943. Most of the other countries did not need to go so far because their smaller flocks had a better basis of absolute chicken feed (feed nothing but chickens will eat, i. e., waste and worms) on farms — and depended less on commercial grain and protein concentrates. In 1942 the German flock was curtailed only 13 per cent. The pig herds met the same fate. Denmark had in October, 1939, a total of 3,192,000 pigs. By March, 1942, there were only 1,267,000 pigs left, but by January, 1944, with an expanded domestic potato and grain crop, the herd had increased to 2,291,000. The Dutch pig herd was cut from 1.6 million prewar to 498,000 in 1943.

The German pig herd was reduced to two-thirds of the prewar average in 1942 but was increasing substantially in the latter part of 1943, to be curtailed later owing to a shortage of potatoes. In France the prewar staple of 7,000,000 swine had fallen slightly below 5,000,000 in 1942. The Swiss curtailed their herd from 880,000 in 1939 to 629,000 in 1943, and further in

1944. In central and southeastern Europe the reduction probably remained within much narrower limits than in Denmark. In general, reports indicate that in 1942 the lowest status was passed and in 1943 recovery of the staple was under way.

Reduction of numbers of chickens and pigs is quickly reparable as soon as more feed becomes available. Both types of animal production permit technically a restoration of numbers and output of products up to or beyond prewar levels within a single year.

The situation differs in the case of the agriculturally all-important cattle herd. In prewar years the Continent had roughly 110,000,000 head of cattle, most of it dairy breeds. More than half the world's milk output came from Europe excluding Russia. This herd in prewar years supplied annually something like 5.5 million tons of beef and veal, and 110-120,000,000 tons of milk. It utilizes a vast acreage of land, which is permanently under grass because it is too wet or too steep or too high to plant to crops. This herd is the most important source of fertilizer so essential to the maintenance of high crop yields on arable land. It takes four years to raise a cow and realise a full milk yield from her. A cow is so valuable that once a small farmer sells several of them and uses the money it takes him years to save enough for replacement. Killing 1,000,000 chickens need not amount to much. Killing 100,000 cows may mean a great deal.

Being fully aware of the vital importance of the cattle herd and of the technical impossibility of restoring losses quickly, all governments as well as the farmers have exhausted all other reserves first to keep the cattle herd (essentially cows) intact so far and so long as possible. All available information points to the conviction that this policy has up to now been successful. The immediate curtailments of hogs and chickens and the reduction of meat rations to consumers put into effect in some countries have successfully checked a later depletion of the cattle stock. Germany maintained her cattle herd through 1943 almost at the 1939 level.

Because the pig herd in 1942 was very low, a small cut had to be made in meat cattle, while the herd of close to 10,000,000 dairy cows seems to have been sustained. In Holland the grass acreage was gradually reduced from 1.3 million hectares in 1939-40 to. 1.1 million hectares in 1943-44 and the herd of milk cows culled from 1.5 to 1.1 million head. In Denmark the cattle stock was reduced from 3.3 million in July, 1939, to 2.9 million in July, 1942, and the cow herd from 1.6 to 1.4 million head in the same period. By July, 1943, the cattle stock had increased to 3,000,000 head. In 1942-43 the French cattle herd reached its prewar count after a loss of roughly 10 per cent in 1940. The number of cows recorded in the 1942 census was not quite up to 1939 status, but it may safely be assumed that in 1942 not all cows were reported.

Military occupation, embittered resistance against the Vichy regime and an ubiquitous black market for all rationed goods are not exactly conducive to perfect honesty in declaring vital farm data. The weight and milk yield of the herd is said to be still below prewar. The Dutch cattle stock at the end of 1942 was reduced only to 2.2 million head compared with 2.6 million in prewar years. Some countries have managed to maintain their livestock herds at prewar numerical strength.

Foreign Crops and Markets published a survey on the changes in livestock in Europe from 1939 to 1942. The tabulation on page 141 shows the result described in detail above.

However, by 1944, the situation as to cattle and hogs had decidedly improved due to the adjustment in crop production and the good grain crop of 1943. Cattle herds in France and in the Danubian basin except Hungary have improved particularly due to the tendency of farmers to hoard grain in this form as long as industrial goods are scarce and meat fetches exorbitant prices in the black market.

So far Europe's agricultural plant has been less seriously depleted than the length of the war and the extent of total economic warfare

Sheep	**Cattle**	**Hogs**	
Year	14 Countries	21 Countries	17 Countries
1939	78,414,000	95,933,000	66,303,000
1942	72,759,000	80,150,000	48,114,000
Percentage	7 per cent decrease	16 per cent	27 per cent

would suggest. Despite many serious handicaps and inconveniences, public policy in every country endeavors to keep agricultural production up or to expand it. In many countries, and in many branches of agriculture, moreover, progress has been made, particularly in the milk-processing industry, in refrigeration of fruits and vegetables, and in dehydration of many foodstuffs.

The Continental area in which agriculture has definitely suffered is the German-occupied sections of Soviet Russia, where thousands of villages and farmsteads were burned to the ground, population and livestock were destroyed or evacuated, railroads torn up and uncounted millions of acres abandoned to weeds.

The war broke the world's communicating system of agricultural areas into several compartments. Nazi Europe is one of them, composed roughly of 350,000,000 or 400,000,000 people. Soviet Russia is the adjacent compartment, separated by the movable partition of the Russo-German Front. While nazi Europe has been cut off by blockade from outside supplies and lives on the output of European agriculture alone, Soviet Russia has been able to obtain food imports from overseas, although at the expense of severe strain. But in spite of these supplementary supplies, Russia's food situation has become critical, because large food surplus areas were captured and upon being retaken were so devastated that they no longer had anything like their normal value.

By November, 1942, when Germany made her greatest expansion into Russian territory, an area was occupied by the enemy containing in prewar years 33 per cent, or 59,000,000 of Russia's population and the following percentages of the country's agricultural output, measured simply by the area's proportional number of animals and acreage of crops:

Beef, Veal and Milk	33 per cent
All Grains	35 per cent
Pork and Bacon	58 per cent
Vegetable Products:	
Cotton and Cottonseed	16 per cent
Flax and Linseed	40 per cent
Potatoes	43 per cent
Sunflower Seed	54 per cent
Hemp and Hempseed	60 per cent
Sugar, Beet tops, Leaves, and Pulp	85 per cent

The prewar population was not all left in the territory when the German army conquered it. The soldiers drafted from there moved out with the army. A substantial force of railroad personnel, MTS station mechanics, tractor and a truck driver, skilled workers and civil servants, was also evacuated. Yet there is evidence that from 4045,000,000 people were left behind and came under German occupation. If this is true, the remaining free Russian territory had to feed from 13-18,000,000 more people than in prewar years, among them some 4-5,000,000 soldiers who needed more calories than the per capita average of the civilian population. To illustrate the point, 8,000,000 German soldiers receive as much meat under army rations as 28,000,000 normal adult consumers, and as much bread as 20,000,000 normal adult consumers. All those civilians who are not heavy workers and on special rations are normal adult consumers. Russia had also to meet the heavier bread requirements of workers in essential war industries.

It is true that until the summer of 1942 Russia's food position was still fairly secure due to large war reserves of grain and the possession of large parts of the areas later lost, but it will be seen that during the winter of 1942-43 it began to deteriorate.

In order to weigh the percentages of lost production previously cited, several qualifications are essential. While the actual loss of population in the occupied area at its maximum was roughly 25 per cent of Russia's population, its equivalent in terms of food requirements was much smaller. The proportion of lost production was on the other hand larger than the figures indicate because they give only the proportion according to livestock numbers and acreage of crops.

For grain, potatoes, and sugar beets, the yield per acre was considerably higher in the lost territory than in other parts of Russia. Thus the loss of 35 per cent grain acreage may have meant a loss of 40 or even more per cent of the grain crop. If the population loss was equivalent perhaps to only 20 per cent or less, then a deficit of grain for 20 per cent or 35,000,000 people has opened up. In a country in which normally 70-80 per cent of all the calories in the diet are derived from cereals and potatoes (in wartime more) this creates a very critical situation. If one calculates the average prewar intake of bread grain per capita of the population including the soldiers as roughly one-fifth of a metric ton, this deficit would amount to 7,000,000 tons of grain. The net loss in potatoes was equivalent to another 2,000,000 tons of grain. A loss of a food supply equivalent to 9,000,000 tons of grain or grain value (i. e., potatoes) and to at least 2,000,000 tons of sugar, plus 300,000 tons of vegetable fats, and 1,000,000 tons of meat is of such magnitude that it cuts deep into every civilian's food supply.

Naturally Russia's agricultural commissar and his administration have not been idle. Grain and oilseed production have expanded in the free areas. In 1941-47, 5,000,000 more acres were sown to grain; in the autumn of 1942 another 6.4 million acres. How much this addition means in yields, and how far the former yields have been sustained is beyond our knowledge.

The Russian food situation became worse — at least for the duration of one crop year — as the Red army succeeded in liberating large occupied areas. By July, 1943, North Caucasus and the region between the Don and Donetz rivers were liberated. From July to December, 1943, all the fertile land east of the Dnieper comprising the eastern half of the Ukraine was freed. By July, 1943, probably 10,000,000 of the original population, and by summer 1944 perhaps several million more people were repatriated. Russia's food supplies are seriously burdened because she is obliged to feed these repatriates.

Up to the date of this writing there is no other area in Europe, or in the world, where agriculture has suffered destruction of physical production assets similar to that along the Russian border and areas of penetration by the invading German armies. Nowhere else has such deliberately planned, and resourcefully executed paralyzation of all the secondary services so essential to the proper functioning of agriculture taken place. Only the eastern part of Poland has witnessed such wanton devastation of cities and farms alike, but there the curse of war passed only once, and years of reconstruction have repaired a good deal.

But the liberated Russian territory experienced a repeated sweep of destruction. When the Red army retreated it applied the torch to elevators and tool sheds, villages and factories. The "scorched earth" policy was supposed to defeat the German armies and to frustrate nazi plans to utilize the Russian west as a source of grain exports. Nearly all the 36 per cent of Russia's tractors, the 37 per cent of her combine harvester-thresher machines, and the 42 per cent of all trucks of this area were either destroyed, disabled by removing vital parts, or moved out. Much worse was the destruction of elevator machinery, creameries, sugar factories, bins, barns, and living quarters on kolkhoz and sovkhos farms. The greater part of livestock was slaughtered and lost, a minor part removed or driven into the forests.

As the German army slowly retreated it applied with greater thoroughness the same strategy of feeding scorched earth to the enemy Sappers and engineer troops were trained in a special demolition school in a Donbas city. Even if no second systematic destruction had been organized and executed, a blighted area would nevertheless have remained. The only great front-line artillery duels in Europe were taking place on the Eastern Front. This is stationary warfare, forced upon the German army by the magnificent staying power of the Russian army. It moves slowly, and hence it is extremely destructive to all the permanent improvements on land, to bridges and viaducts, power dams and power plants, and to every building in industrial areas or on farms. Dynamite and torch, fire and explosion carried by bombs, mines, shells and flame throwers do the work.

The Russian people left behind in this devastation are decimated by famine and disease, by acts of warfare and reprisal, but most of all by a selective draft by the German forces of occupation that sends to Germany and elsewhere the most useful elements, male and female, old enough to work. Whatever livestock can be moved out is rounded up and shipped back. A Reuters dispatch of October 3, 1943, stated briefly that according to official German high command figures triumphantly broadcast from Berlin, German troops withdrawing to the Dnieper's western banks took with them 106,000 head of cattle, 110,000 sheep, 60,000 horses, and 62,000 Russian civilians. The appalling thing is the smallness of these boastful figures. They illustrate how little was left to take alive as booty.

The fields are overgrown with weeds since there is neither manpower nor animal draft power nor tractor power to plow and sow them, nor is there seed to sow, nor food for those who would begin the cultivation, nor will there be for a long time to come. Considering the millions of liberated people, Russia's grain deficit is such that, if shipping permitted, at least 3,000,000 or 4,000,000 tons should be imported, in order to keep the privations of the people within bearable limits and to create some small reserves. Under the given circumstances, whatever the available ships and the port and railroad facilities from Murmansk and Bandar Shahpur (Gulf of Persia) and Vladivostok can bring beyond the military priority equipment is pitifully little. Moreover, most of the food imported is reserved for the armed forces.

Just how serious Russia's food situation is depends ultimately on the size of the stocks of grain available. There is little doubt that Russia's grain reserves have been much larger than is generally realised and that its management and partial disposal has been one of the most vital parts of the country's food strategy. Even without any specific knowledge of the procedures during the war, the foreign observer could guess that Russia did not allow granaries to be caught empty when the war broke.

The government which nearly completed its third five-year plan, which had for twenty years struggled with the food situation, which had passed through years of stark famine, and which proved to be in such excellent military shape when its territory was invaded, could not possibly have ignored the most basic requirements of food reserves, particularly not in view of the accumulation of very sizable stocks in Germany since 1937.

Soviet Russia had rationed bread for the nonfarm population from 1933 through 1935. During those years a special reserve carry-over of 5-6,000,000 tons of grain beyond the 4-5,000,000 tons maximum stock requirements were accumulated, so that in 1936 the country's total stock on June 30 was probably in the neighbourhood of 10,000,000 tons. During the following years the crops fluctuated greatly because of weather. The years 1937 and 1941 brought bumper crops; 1938 a poor crop; and 1939, 1940, and 1942 average ones. The crop of 1937 probably was closer to 95-98,000,000 tons than the 120,000,000 tons officially estimated. Of the 1942 crop, about one-fourth was probably lost in the process of harvesting. The government collected grain from the farmers and added to its special reserves as follows:

Year	Government Collection	Addition to Reserve	Total Reserve
	(in million metric tons)		
1937	32	5	10-11
1938	18	-	10-11
1939	39	8	18-19
1940	39	8	26-27
1941	?	?	26-27

In 1941 some 9,000,000 tons of the grain reserve were lost to the enemy, since more than a third of all the storage was located in the invaded area. Thus at the end of 1941 there were still 18,000,000 tons of stocks left in the hands of the Soviet government, and the farmers held 5,000,000 tons. In 1942, 7,000,000 tons of the reserves had to be used to supplement the new crop. Yet there were probably still 11,000,000 tons held by the government and some additional grain held by the farmers. The year 1943 saw a vast deficit in grain when occupied areas, needing seed as well as food for its people, were liberated. Moreover, the synthetic rubber industry and the manufacture of explosives required 3,000,000 tons of grain for industrial alcohol a year. The possible grain reserve of 11,000,000 tons would probably be absorbed completely if the city bread rations were not kept as low as they are.

From these few data which are the best available estimates by American experts, and which naturally contain a considerable margin of error, the reader can at least grasp that the immeasurably greater leverage lies in the intelligent adjustment of food and feed rations, and in using stocks, rather than in importation.

The other great adjustment lies in the reduction of the livestock herd. According to American estimates, the herd on free Russian territory excluding the front zone has declined below the 1938 figures as follows: cattle, 15 per cent; sheep, 20 per cent; farm horses and pigs, 25 per cent. The decline in farm horses is not a gain for the feed requirements, because the decline expresses simply the drafting of horses for the army. The total number of nonmilitary and military horses is probably larger than in 1938, and probably consumes as much or more grain because army horses are fed better rations.

Since, together with the Chinese, the Russian people are among the world's hardiest races, and have demonstrated great physical endurance and mental stamina to stand up under grinding hardship and scanty food supplies, the Soviet Union is determined to shorten the war by sacrificing even dietary necessities, by importing cannon rather than butter to paraphrase General Goring's statement. Russia gets along somehow with her short supplies by tightening the belt, by feeding fewer animals, by scraping together and stretching out a considerable net gain, and by importing partly on American lendlease account, partly from British sources and elsewhere, substantial amounts of fats, meats, dried milk, and sugar, and relatively small amounts of flour and grain. How black the prospect for an early restoration of agricultural production in the liberated areas must be is best illustrated by the government's decision to let only very few of the evacuated people return to the Eastern Ukrainey and White Russia. It would probably take too many essential war materials and in general sap more strength from the war effort than it would add to it by greater food production on devastated areas once they could be restored. In the long run this decision may also be advantageous for the sake of a better distribution of the population with more people living in Siberia, something that ever since Peter the Great every Russian regime has tried to achieve.

In the fall of 1943 Russia made a successful sowing campaign in the Kuban district. In the eastern Ukraine at least some grain was sown. In the western Ukraine the Germans left sown fields behind. With additional spring sowing under way there is at least enough sown acreage to promise a moderate grain crop — provided adverse weather does not cause a failure. Dispatches of early March, 1944, from Moscow indicated that airplane and ordnance factories are putting great pressure on the production of spare parts of tractors and combines to insure the spring farming campaign. With nearly all of her territory liberated, a much more extensive sowing in the fall of 1944 may initiate an era of greater grain supplies for the 1945-46 food-consumption year. During late spring 1944 the Russian civilian population probably passed the worst months of the food situation of this war. Bread and potatoes will probably become more available. To rebuild the livestock herds will take several years.

Areas of the continent have probably improved their agricultural production capacity, which ones have maintained or only slightly lost ground, and where various degrees of serious depletion have occurred from September, 1939, through early spring 1944.

The chart is self-explanatory. Naturally it does not tell where food is scarcest, because in a country like Belgium or Switzerland, agricultural crop production has been substantially improved during the war and yet with the absence of former heavy food imports the provisions, for the urban population in particular, may be far below the prewar supply. Chapter VI will consider this aspect of the economic effects of the war more closely, since the food shortage constitutes the demand for agricultural exports from overseas.

For the time being we continue our review of the effects of war upon world agriculture up to the winter of 1943-44. Since we have dealt so far with the two huge continental blocks of nazi Europe and Soviet Russia, one of which suffers from the blockade and the other from repeated scorching of the earth and destruction in combat zones, and both of which are making both ends meet mainly by underconsumption and adjustments in diets and utilization of food, it serves our purpose best to survey briefly the other vast but tightly closed compartments of the food realm — those of China and of the Japanese controlled territory.

12

The Environment of Agriculture

SOCIOLOGY OF AGRICULTURE

The tack we have taken in developing two chapters on recent research in the sociology of agriculture has been to make an (admittedly arbitrary) dichotomization of research on the basis of whether it focuses on the "internal" dynamics and characteristics of agriculture on one hand, or on the "external environment" of agriculture on the other. This chapter, in focusing on the "environment" of agriculture, employs this expression both metaphorically and literally. More specifically, we begin by focusing on four selected aspects of the metaphorical or social "environment" of agriculture—technological change, the sociology of agricultural research, the impacts of farm structural change on rural communities, and the farm crisis. We then turn to several areas of inquiry in which agriculture is explored in terms of the natural or physical environment. As noted earlier, the dual theme of the ecological consequences of agriculture and the role of environmental factors in affecting the structure of agriculture has been one of two key lines of theoretical development—the other being neo-Marxist and neo-Weberian-derived versions of the political economy of agriculture—that have spearheaded the "new sociology of agriculture" since the early and mid-1970s.

TECHNOLOGICAL CHANGE ON AGRICULTURE

Technological change is among a small handful of topics in what is now referred to as the sociology of agriculture—

along with part-time farming and analysis of the farm family—that have received attention during all three major phases in rural sociology. From the 1930s through the 1950s there were several major studies by sociologists of the impacts of technological change, most of which centreed on the rapid mechanization of southern agriculture that began during the Great Depression. The concern of many of the early rural sociologists with the socioeconomic consequences of technological change, however, largely lapsed during the succeeding phase of rural sociology in which the diffusion-adoption perspective was dominant. In the era of the "new sociology of agriculture," we again see a rekindling of concern with the social consequences of technological change. Moreover, interest in the consequences of technological change has become expressed in more social structural terms and, as demonstrated later, has become more closely integrated with analyses of the genesis of new agricultural technologies.

Several other important observations can be made on the rural sociological literature on technological change in agriculture over the past decade. First, mechanization has received the lion's share of attention; technological change in the forms of improved varieties of crops and breeds of livestock or of agricultural chemicals has been the focus of only a few major ex post studies. Second, as Berardi has observed, there are many problems in decomposing the effects of technological change on farm structure from the effects of other variables such as public policy or changing wage rates; therefore, rural sociologists have often relied heavily on the primary research of others (e.g., agricultural economists and geographers with complementary expertise in assessing technological impacts) when developing comprehensive assessments of the socioeconomic impacts of technological change.

Third, the conceptualization of technological change in agriculture by rural sociologists has been heavily influenced by the work of the agricultural economist, Willard Cochrane, who developed a theory of the "treadmill of technology" in the late 1950s. Briefly, Cochrane's theory of technological change builds on a key observation from rural sociological research on the diffusion-adoption of innovations-that large farmers who are entrepreneurial and nonrisk-averse tend to be the first to adopt new agricultural technologies. By adopting early, these farmers enjoy "innovators' rents" because of their ability to reduce their average per unit production costs.

Cochrane posits that after these innovators and early adopters have employed the new agricultural technology, aggregate output of the particular commodity involved will begin to increase and prices will decline disproportionately (because agricultural commodities tend to have a low price elasticity of demand). Declining prices then begin to force nonadopters to utilize the technology. These subsequent adopters gain very little from the technology; they adopt it merely to be able to stay in business. Slow adopters or nonadopters will tend to be forced out of agriculture because they can no longer compete due to their high average costs of production. Cochrane considers the technological change process to be a "treadmill" in that the majority of farmers, who receive little or no benefit from the technology, are nonetheless forced to continue to adopt new agricultural technologies in order to stay in business.

Cochrane has also observed that the benefits of technological change tend over the long term to be realised in the form of rising land values, particularly if there are commodity programmes in effect to cushion a commodity sector from declining crop prices caused by increases in output. The benefits of technological change (innovators' rents from early adoption) tend to be reinvested in farm land, causing increases in land prices and benefiting large landowners over smaller property owners and tenants. These effects, in Cochrane's view, are heightened when commodity programmes support crop prices, thereby increasing the level of innovators' rents and intensifying the extent to which these benefits are capitalized in asset values.

Cochrane's theory of the treadmill of technology has been widely accepted in recent rural sociological work as an orienting perspective because it has effectively linked knowledge from diverse origins (rural sociological research on diffusion-adoption, agricultural economics research on technological change, and so on). Cochrane's theory, however, has its limits—in particular, its inapplicability to industrial agriculture, as noted by LeVeen. Moreover, the notion of the treadmill of technology gives little attention to the origins of technology, that is, to why technologies that have displaced labour or benefited larger operators over smaller ones have been developed, a topic that has recently come to be a major research issue for those working in the sociology of agricultural science. Nonetheless, with the exception of sociologists of agriculture who have focused on industrial agriculture, rural sociologists have yet to develop a distinctly sociological theory of technological change and its socioeconomic consequences.

While rural sociologists have not yet developed a comprehensive and distinctly sociological theory of technological change and its consequences applicable to both family farming and industrial agriculture, the new sociology of agriculture as applied to technological change has made some significant advances. Several of the most important of these advances are summarized below.

Among the most significant research papers on the topic of technological change in agriculture were two published in the late 1970s that dealt with limitations of the classical diffusion-adoption model. The first, by Pampel and van Es, was an otherwise traditional adoption study that included, in addition to "commercial-profitable" innovations that were the major focus of previous diffusion-adoption studies, a set of "environmental innovations" such as contour farming, terracing, sod waterways, and use of rotation cover and reduced tillage. Pampel and van Es discovered that the correlates of adoption of environmental innovations were far different than those of commercial innovations. For example, while large-scale farmers were observed to be the earliest adopters of commercial technologies, there was no significant relationship between farm size and the adoption of environmental innovations.

They concluded their study by cautioning rural sociologists that the diffusion-adoption tradition, because it had ignored dimensions of innovative behaviour such as the adoption of environmental-conservation technologies, "may have provided within rural sociology a field of knowledge with a narrow empirical foundation on which to base its generalizations". They also emphasized that given the tendency of most environmental innovations to be unprofitable, it was not clear that voluntary strategies to achieve soil conservation and related environmental goals would be successful.

Writing two years after the Pampel and van Es study, Goss focused more systematically on the limitations of diffusion-adoption research. The principal failings of diffusion-adoption research, in Goss' view, were

1. That it was uncritical, if not promotional, toward patterns of technological change that may be detrimental to the interests of society and/or farmers,
2. That the focusing of research effort on the adoption component of the technology change process had caused rural sociologists to ignore the consequences of technological change for various groups in society. Goss summarized the early 1970s literature critical of the Green Revolution and pointed out that the Green Revolution had been initiated with the same promotional orientation toward technology that had predominated during the heyday of the diffusion adoption tradition in U.S. rural sociology. He noted that many social scientists who had once worked within the diffusion-adoption tradition in international agricultural activities had begun to recognize the limits of the adoption paradigm. He argued, however, that this point had not been well recognized in domestic rural sociology and that rural sociology must begin to develop a more comprehensive programme of research on the consequences of technological change in agriculture.

In a manner similar to Goss' exegesis of the Green Revolution literature in terms of the impacts of technological change on social groups and classes in agriculture, Stockdale, in an influential article, raised comparable issues in terms of the ecological aspects of modern agricultural technology. Noting that the contribution of technological change to American agricultural productivity had often been held up as a model for other nations to emulate, Stockdale argued that social scientists had tended to ignore the adverse ecological impacts of energy-intensive technology in the United States and elsewhere. Like Goss, Stockdale called for an expanded programme of research on technological change in agriculture that would emphasize the social and ecological consequences of new agricultural technologies.

Nonetheless, Friedland and his associates have demonstrated that the technological change process in industrial agriculture generally is heavily influenced by considerations of the cost of, access to, and control over labour and by the relationships of grower organizations with public and private components of the agricultural research system. They have also demonstrated that the nature of the technological change process and its consequences depend upon the structure of the commodity system. For example, in the aftermath of the termination of the bracero programme in 1964, a mechanical tomato harvester technology was widely adopted in California, displacing three-quarters of California tomato producers and the bulk of their agricultural wage labour forces. In lettuce, however, grower-shippers were able to utilize the threat of lettuce harvest mechanization to discipline the labour force and to avoid the costly process of harvest mechanization.

A further component of the recent rural sociological literature on technological change, one to which Friedland and Friedland and associates have made major contributions, is social impact assessment of new agricultural technologies. Social impact assessments of new agricultural technologies are a distinctive type of research in that they often tend to rely primarily on secondary data or on the results from the published literature rather than on primary research and data collection; this is particularly the case if the assessment must be prepared over a short period of time.

Other distinctive aspects of social impact assessment of new agricultural technologies are, first, that the assessment is ex ante (rather than ex post, as in most previous rural sociological research on technological change) and, second, that the orientation toward these emerging technologies tends to be questioning and skeptical, rather than accepting and promotional. Rural sociologists who have conducted social impact assessments of new technologies have approached their subject matter from the vantage point that most new technologies tend to involve an unequal distribution of costs and benefits and that advance projections of these patterns of the distribution of costs and benefits can enable social scientists and affected groups to redress these distributional impacts more effectively.

An example of the type of knowledge base that has been developed, in part, as a building block for social-impact assessment research is Berardi's comprehensive review of the impacts of agricultural mechanization. Berardi reviewed the results of over 180 studies of agricultural mechanization in the United States and elsewhere and summarized the major empirical regularities in studies of a variety of types of mechanization in a diversity of commodity systems. She concluded by pointing out that the rich literature on agricultural mechanization—most of it prepared by nonsociologists—is sufficient so that sociologists may contemplate conducting rigorous ex ante studies of the impacts of new mechanical technologies in agriculture. Unfortunately, as noted earlier, sociological knowledge on the socioeconomic impacts of mechanization is far greater than that on "biochemical" technologies (plant varieties, fertilizers, insecticides, herbicides, and so on). There have only recently been sociological studies that have delivered on Stockdale's and Berardi's admonitions in the area of biochemical technologies in plant and animal agriculture. It is likely, however, that the great attention that has already been paid by rural sociologists to emerging biotechnologies in agriculture will result in a sustained research programme into the socioeconomic consequences of these new technologies.

It should be noted in closing, however, that while there is great merit in inventorying the results of existing studies on various facets of technological change—as, for example, Berardi has done—sociologists should exercise caution, as technological change should not be assumed to have invariant consequences across time and space. In particular, as noted above, there have been some major recent changes in the structure of U.S. agriculture that one might expect will have significant effects on the technological change process and its consequences.

For example, instead of technological change tending to result in the decline of small farms, as it did during the bulk of the post-World War II period, it may be quite likely that new agricultural technologies such as biotechnology-derived plant varieties, growth regulators, and bovine growth hormone may tend to result in the disproportionate demise of medium-sized, full-time family farmers. Thus future rural sociological research on technological change in agriculture must be cognizant of ongoing changes in the structure of agriculture and how these changes affect the types of agricultural technologies that are employed, the processes of technological change, and their consequences.

Agriculture and the Rural Community

Perhaps in no other area of the "new sociology of agriculture" is there so much apparent continuity with the pre-Korean War tradition of rural sociological research as in research on farm structure and rural communities. The early rural sociologists primarily conceived of agriculture as one of the major institutional ensembles affecting rural community life and, especially during the 1930s, devoted much of their research effort relating to agriculture to understanding the implications of agricultural structural changes for rural community life and viability. The conceptualization of agriculture and rural community interactions and the methodologies for exploring these interactions in pre-Korean War and post-1970 rural sociological research, however, exhibit dramatic differences. Whereas most of the early rural sociologists tended to be less interested in agriculture per se and were principally concerned with understanding rural community institutions, contemporary researchers have tended to be more interested in structural change in agriculture and have deemphasized understanding the totality of the fabric of rural social life.

The research tools and data sources of these two generations of rural sociological researchers are also dramatically different. The early rural sociologists relied on rural social surveys, while the bulk of contemporary research is based on county-level census data. A further difference between the early and contemporary rural sociologists of agriculture in their work relating to rural communities concerns the pivot of the latter studies around Goldschmidt's "As You Sow".

This is somewhat ironic in that Goldschmidt's study was anthropological and qualitative, whereas the research conducted since 1970 that has drawn on Goldschmidt's work has largely been quantitative and has involved far less "anthropological" detail on rural communities than did Goldschmidt's. Briefly, Goldschmidt's thesis was that there are differences in the quality of rural community life depending upon the social organization of agriculture and its attendant occupational structure. Goldschmidt observed that the quality of rural life in two carefully matched California communities in the San Joaquin Valley differed dramatically and that these differences were largely accounted for by the organization of agriculture.

He argued that Dinuba, a community surrounded by "family" farms, had a greater number of businesses, a larger retail sales volume, a higher per capita income, and a greater diversity of social, educational, recreational, and cultural institutions than did Arvin, a community surrounded by large, industrial, corporate farms. These differences were principally accounted for by an intervening variable—community occupational composition—that was seen to be strongly influenced by the social organization of agriculture and which directly contributed to the dramatic variations in indicators of the quality of life. Goldschmidt reported that the high ratio of labourers to other occupational groups in Arvin accounted for its low level of community viability relative to Dinuba. From these observations Goldschmidt inferred that the rise of large-scale corporate agriculture then underway in California and elsewhere would have adverse implications for rural communities.

Goldschmidt's study received essentially no further attention until it was referred to in a 1968 hearing on corporation farming before the Senate Select Committee on Small Business. LaRose and Sonka and Heady soon reported further data consistent with the Goldschmidt study. In addition to Rodefeld's and Heffernan's efforts to resurrect Goldschmidt's work in rural sociological circles, there was a major research effort by the Small Farm Viability Project in California, drawing largely on Isao Fujimoto's unpublished research, which presented a larger restudy of the Goldschmidt thesis. The results largely supported Goldschmidt's findings of over three decades earlier. Flora and Conboy also reported results consistent with the thesis.

Since that time there have been numerous studies in the new sociology of agriculture tradition that have explored the Goldschmidt thesis in a number of contexts.

Several observations can be made on the conduct of recent research in the Goldschmidt tradition, what we might call "neoGoldschmidt" research. First, it must be stressed that contemporary rural sociological researchers have taken considerable liberties with Goldschmidt's formulations. As noted earlier, Goldschmidt's method was a comparative anthropological case study of two communities and eschewed quantitative analysis of the sort that has predominated in the subsequent rural sociological literature. Also, Goldschmidt's analysis focused on several aspects of the property relations and control over labour, such as absentee ownership and use of ethnic minority labourers, rather than farm size, which has been the most important farm structural indicator in the neo-Goldschmidt tradition. In that regard it is important to stress that the "family" farms of Dinuba were by no means "small" farms; in fact, the family farms of Dinuba were generally larger in gross sales terms than most large farms in the Midwest today.

Second, neo- Goldschmidt research has largely focused on a single direction of causality—the impact of agricultural structure on rural communities—and has largely ignored other types of interrelations between agriculture and rural communities. Third, most studies have been based on Census of Agriculture data and have employed countylevel data as a proxy for community characteristics, though some researchers have utilized survey data to infer community quality of life consequences of changing agricultural structures. Fourth, despite obvious major regional variations in farm structures and in rural communities and their economies, there has, until very recently, been a virtual lack of a comparative perspective on farm and community structures.

Goldschmidt and Harris and Gilbert are partial exceptions to this generalization, since they have done research using states as the unit of analysis, though this procedure compounds the aggregation bias problems. Very recently, nonetheless, there has appeared a series of regionally based studies of the effects of farm structure on rural community viability and quality of life commissioned by the Office of Technology Assessment. These studies are Buttel et al., Skees and Swanson, Flora and Flora, MacCannell, and van Es et al. as published in abridged form in Swanson. These five studies were very briefly summarized in the main report. Together they indicate strongly that there are pronounced regional variations in how changing farm structures affect rural communities.

The impacts of farm structural changes on rural communities are generally most pronounced in the Great

Plains and industrial agriculture states less so in the South and Midwest, and very little in the Northeast. There is evidence that the degree to which farm structure affects rural communities is directly associated with the dependence of the rural community on agriculture.

The general model pursued in these neo- Goldschmidt studies has been to estimate coefficients for a causal chain in which farm structural change indicators are seen to affect the composition of the farm population and the size and composition of the rural population. These clusters of variables are then seen to affect indicators of the quality of life in rural-agricultural communities. This general model is sometimes supplemented by technological factors as antecedents of farm structure, since technological change is considered to be one of the major factors influencing farm structural change. More recently, studies in this genre have gone beyond the cross-sectional census data designs that predominated from the mid-1970s to the early 1980s and have included measures of change, where appropriate, in variables in the model. One limit to studies that use census and other data based on single-year assessments of farm income is that net farm incomes are highly variable over the short-run; for example, as many as 20 percent of farms with annual sales of over $250,000 had poverty-level incomes in 1986.

Heffernan has effectively summarized the general configuration of results in the neo-Goldschmidt literature by noting that "all relevant research to date suggests that a corporate type of agriculture results in a reduction in the quality of life for at least some people, especially the hired workers in rural communities." As such, the results continue to lend support to the Goldschmidt thesis. But there have recently been several expressions of dissatisfaction with Goldschmidt's perspective and with the generalizations that have been made from research in this tradition. A major theoretical concern is that Goldschmidt and many researchers working in this tradition have taken large-scale and corporate agriculture to be basically synonymous when, in fact, not all large-scale farms are corporate or industrial-type farms. Much more important than industrial-type farms in farm numbers and shares of sales, assets, and profits are larger-than-family farms.

Empirically, as Goldschmidt-style research has been done in the Midwest and other traditional family farming areas, there have been admonitions that large-scale agriculture, especially of a nonindustrial character, does not necessarily lead to adverse impacts on rural communities, though some studies note that increased scale of production could have detrimental impacts on community services.

A further area of dissatisfaction with research on farm structure and rural communities is that the treatment of farm structure in this literature has generally failed to incorporate new concepts relating to agricultural dualism. Typically farm size or scale is operationalized as a single variable representing a central tendency measure of farm size, which was a suitable procedure when there was a singular process of differentiation among farms. But with the emerging dualism in U.S. farm structure since the late 1970s, there has been little change in mean acres per farm since the 1969 census. This measure of central tendency thus ignores the processes of dualism that have contributed to increased concentration with apparent stability in average acres per farm since the late 1960s.

The neo- Goldschmidt conceptualization of an agricultural structure in which particular communities can be characterized along a continuum of farm sizes ranging from small to large has also been questioned by Wimberley who, using multiple indicators from county-level Census of Agriculture data and a longitudinal design, found that there are instead three continua relating to degree of "corporate-commercial" agriculture, "largefarm-area" agriculture, and "small-farm" agriculture. Using Wimberley's measures and including sources of off-farm employment as well, Reif found that large-farm-area agriculture and core-industry structures were related both to family poverty and income inequality in counties.

Thus, while studies of the impacts of changing farm structures on rural communities have been a productive area of research in the new sociology agriculture, a number of research frontiers for the future are suggested by recent criticism. There is a need to diversify the methodological approach, consider alternative directions of causality, introduce more sophistication in terms of the measurement of change in farm structure, and expand recent efforts at comparative regional and interdisciplinary research. The work of Vail represents an encouraging thrust in this direction. Vail has conducted extensive research on the relationships between agriculture and rural communities in Maine in which a variety of factors of demonstrated importance—the reciprocal relations between agriculture and rural communities, the role of an increasingly dualistic agrarian structure, nonfarm aspects of agriculture, and the historically specific nature of farm-community interactions—have been incorporated into a single study.

The Farm Crisis Of The 1980s

The notion that American agriculture is no longer isolated from the remainder of the political economy was driven home clearly by the farm crisis of the 1980s. This crisis was brought by the conjunction of a number of factors in the 1970s, including: accelerated demand for grain on world markets, public policies that encouraged massive capital investments and expansion of production, and a political-economic context that encouraged land speculation. The farm crisis emerged in the 1980s for a number of reasons. Global agricultural export markets shrank, in large part due to the global recession.

Farm commodity prices and ultimately net farm incomes declined due to the glutting of world grain markets. Reagan administration fiscal and monetary policies led to high real interest rates, which were especially detrimental to agriculture because of its interest rate sensitivity; accordingly, debt-asset ratios among farmers increased rapidly. Third World agricultural export competitors because of the global debt crisis and these countries' need to increase exports in order to repay foreign loans. Further, federal fiscal austerity eroded the federal government's ability to maintain farm incomes through commodity programmes. The result was a massive decapitalization of American agriculture, especially in the Midwest and Great Plains, as farmland prices in the states most affected by the crisis declined by over 50 percent.

This crisis attracted a good deal of attention by the mass media and by agricultural economists. It was not, however, until the middle of the decade that much was published about the farm crisis by rural sociologists. Most of this research can be divided into four categories:

1. Neo-Marxian (and related neo-Weberian) analyses of the continuing penetration of capitalism into and the role of the state in agriculture,
2. Analyses of the differences in the characteristics between farmers that were and were not experiencing "financial stress,"
3. Studies of the effects of farm financial crisis and failure on the farm families whose enterprises were affected,
4. Studies of public reactions to the farm crisis.

The neo-Marxian (and neo-Weberian) approaches have largely focused on understanding continuing cycles of crises in agriculture under capitalism. Examples of this genre specifically focused on the crisis of the 1980s include Berlan, Bonanno, Mooney, McIntosh and Zey-Ferrell, Zey-Ferrell and McIntosh, Goodman and Redclift, and Buttel.

Studies in the second category, analyses of the characteristics of farms with various levels of "farm financial stress," deal more with the effects of the political and economic environment of agriculture on farms of different characteristics. These studies have produced inconsistent results due to regional, ethnic, and other variations within the farming population. For example, farm-size variables were found to be directly correlated with financial difficulty on Iowa farms and unrelated for Texas and North Dakota farms. Yet aggregate national data revealed that larger farms were more profitable, a relationship that obscured the underlying failure of many middlesize farms. Farm families with fewer years of farming and younger ages were affected more severely. However, Murdock et al. found the cohort that had started farming between the end of World War II and the mid1960s also had higher rates of financial difficulties, suggesting a life-cycle effect for those farm families bringing another generation into the farm. The literature does, however, show agreement on the relationship between other variables and financial crisis. For example, higher debt-to-asset ratios brought by low commodity prices in a time of deflation of asset values were associated with financial problems.

The third category, analysis of the effects of the farm financial crisis and failure on the farm families whose enterprises were affected directly, has largely been the work of William and Judith Heffernan, who have conducted in-depth interviews among Missouri farm families who were in severe financial crisis or had been displaced from agriculture. These families experienced a host of difficulties beyond the simple loss of their farms and the disorientation that accompanies the loss of a valued way of life. Among their major results are the following: Many farm families felt socially isolated and were not well-prepared for participation in nonfarm labour markets. Local employment opportunities were often unavailable, and tax laws devastated some families economically after they had attempted to leave farming with some resources with which to start a new life. While the long-term effects of farm failure for these families were not yet known, it was clear that some families were adapting to their new situations successfully while others were at risk of suffering long-term poverty. Graham's study of displaced farm families in New York State bears out these findings. This category of studies has implications for the mutual effects of agriculture and its environment as the displaced farm families affect local labour markets and community social structures.

The final category of research on the 1980s farm crisis, on public reactions to the crisis, is best represented by Lyson's research on public apathy. Drawing on theories of public opinion formation, Lyson demonstrates that several factors led to a surprisingly high degree of apathy by nonfarm people toward the farm crisis, despite the massive media attention the crisis attracted. He notes that there has been a trend toward farmers and consumers having very little direct contact and that the relationship between food prices and farm income has become concealed. Most importantly, Americans have come to view "bad news" from the farm sector—the cost-price squeeze, high debt loads, and the loss of family farms—as being normal. Thus the American public, though "interested and concerned" about the economic woes of farmers, exhibited apathy toward the farm crisis.

A major neglected area in the rural sociology literature on the farm crisis to date is studies of the effects of the farm crisis on rural communities. This topic pertains, in particular, to how conditions in agriculture affect its environment, especially at the local level. Financial losses to local agribusiness firms when farms fail, losses of farm families in local rural market areas, declining property tax revenues due to deflated farm asset values, and other problems associated with the declining prosperity of agriculture have likely had severe effects on local services and on the local social fabric.

THE CASE OF SOIL EROSION AND CONSERVATION

Environmental problems are by no means new to American agriculture. The Dustbowl episode of the 1930s was but one of several historical instances in which concern has been raised about the environmental performance of U.S. farmers. Nonetheless, current controversies about the management of agricultural resources in the United States date largely from the rise of the environmental movement in the 1960s and have included issues relating to soil erosion, the use of agricultural pesticides, energy and water consumption, and the like. The soil conservation implication of American agriculture has remained among the most longstanding agricultural-environmental issues of concern to rural sociologists and is given principal emphasis in the section that follows.

As noted earlier, the majority of U.S. farms are family proprietorships in which there is household or family ownership and management of the land and water resources. Accordingly, the logical point of departure in understanding the soil erosion problem in agriculture has been to conduct micro-level research on the individual, farm firm, and ecological factors associated with soil erosion rates or the utilization of soil conservation technologies. This has been a productive line of research, but it has also become a controversial one, with the major controversies centreing around theoretical approach, the relationships between micro and macro levels of analysis, and the public policy inferences that should be drawn from social science research.

Soil Erosion and Conservation

One of the most crucial aspects of soil erosion research is the very conceptualization of why soil erosion is a socioeconomic problem. Traditionally this has not been an issue, since sustained high levels of soil loss were presumed to irreversibly degrade the productivity of agricultural resources. Accordingly, if the soil erosion problem is conceptualized as medium to long-term land productivity decline detrimental to the interests of farmers, it would follow that conservation should be achieved through voluntary farmer compliance (rather than through mandatory regulation) based on the logic of the long-term interests of farmers. Historically, federal and state government policy thus has emphasized education of farmers and modest levels of financial inducements (such as Agricultural Stabilization and Conservation Service cost-sharing programmes) to lever farmer decision making. But it has been increasingly recognized that the proportion of prime U.S. farmland that stands to be irrevocably degraded by soil erosion is relatively small and that the "off-site" costs of soil erosion—the impacts of soil erosion and run-off on water and land resources of a farmer's parcel are very significant and may very well be in excess of the on-site costs.

These alternative conceptualizations of the soil erosion problem —productivity loss versus destruction of off-site water and land resources—have very profound implications for appropriate types of social science research and for public policy. To the degree that soil erosion primarily results in land productivity losses, research should focus on individual- and farm-level factors that influence resource management, and public policy should incorporate these research findings in order to develop educational and incentive programmes to achieve soil conservation through farmer self-interest. To the degree that soil erosion is primarily a problem because of off-site costs, farmers cannot be expected to conserve because the long-term productivity benefits of conservation are small, and research at a more macro level would be most appropriate.

Debate over the appropriate kind of analysis in soil conservation research has been closely related to, and in many respects was stimulated by, debate over the adoption-diffusion approach. Pampel and van Es as noted earlier, reported that the correlates of adoption of conservation technologies tended to be different from those of commercial nonconservation technologies. They also suggested that effective conservation practices will tend not to be profitable for farmers, so that to achieve the public interest in soil conservation voluntary compliance among farmers may be insufficient. The issues raised by Pampel and van Es have continued to pervade the erosion and conservation literature. Many researchers have conducted their research by seeking to revise the microsociological diffusion of innovations approach, while others have argued that the constraints to soil conservation behaviour tend to be more macro in nature and that the diffusion of innovations approach will be inadequate.

Prior to 1970 soil and water conservation technologies and practices were generally understood to be a diverse collection of cropping patterns crop rotations, contour planting, strip cropping, cover cropping, sod waterways, filter strips and physical and biomass structures terraces, sediment retention basins, diversion channels, animal waste structures, and hedge rows. Since that time, however, soil and water conservation has, for the majority of researchers and policymakers, come to be largely conterminous with the adoption and use of no-till and related conservation and reduced tillage equipment, which enables farmers to leave a mulch layer of residue on the soil to reduce runoff and sediment losses.

Two observations can be made with regard to traditional and reduced-tillage equipment approaches to soil and water conservation in agriculture. First, these two approaches, while by no means incompatible, are nonetheless quite different and will tend to apply most appropriately to different types of farm operations. Traditional cropping practices and structures are often most attractive to smaller, more diversified farmers, while reduced tillage tends to be most attractive to larger, more highly mechanized producers of row crops. Second, reduced tillage, although acknowledged to be effective in reducing erosion in row-crop monocultures, has become somewhat controversial. One controversial aspect of the reduced-tillage approach has been that it requires increased use of herbicides, which may lead to contamination of water and thereby negate some of the runoff reduction benefits of reduced tillage. Many studies have indicated, moreover, that reduced tillage practices have been adopted by farmers largely because of labour savings rather than because of the ability of the technology to reduce soil erosion.

Indeed, the farmers who have been most likely to adopt no-till and other reduced tillage technologies have been found to be large operators with highly specialized operations involving practices such as continuous cropping and row-crop monocultures. Thus farmer-adopters of such conservation technologies ironically represent the major trends in U.S. farm structure that have been demonstrated to lead to environmental degradation in agriculture. However, farmer perceptions of a threat to soil productivity may lead to adoption of soil-conserving innovations, whereas stewardship related motivations may have little effect on motivations to prevent other kinds of consequences such as off-site pollution.

Sociological knowledge of soil conservation has advanced rapidly, and there are now a number of demonstrated empirical generalizations, mainly at the micro level, but increasingly so at the macro level. One set of factors that has been found to affect the utilization of soil conservation practices consists of the agroclimatic characteristics of farms and farming regions. The adoption of reduced tillage technologies has been most rapid in regions of the country with the most erodible soils. Also, within a particular region, adoption of soil conservation practices is generally greatest on the most erosion prone lands.

It has also been found that farmer attitudes toward soil erosion their perceptions of the severity of erosion on their farm or of the urgency of controlling erosion have minimal or, at most, modest impacts on conservation behaviours. Farmers thus tend to utilize conservation practices independent of their perceptions of the extent of erosion on their farms. Farmer attitudes regarding stewardship obligations toward the land have been found to have significant but modest positive impacts on the adoption of conservation practices.

As noted earlier, most studies have found that indicators of farm size and income gross farm sales, net worth, use of hired labour, acres, etc. are positively related to the adoption of soil conservation practices. Heffernan and Green have also found that farm size is inversely related to the extent of soil erosion, which they attribute more than do most researchers to larger farmers having higher quality, more gently sloped lands. The impact of farm size on conservation practices appears to be greatest for reduced tillage practices than for traditional cropping and structures practices. Evidence indicates that the tendency for larger farmers to adopt conservation practices earlier and more extensively than smaller farmers is reduced when public education and cost-sharing programmes are in effect. It has also consistently been found that educational level of the farmer is positively related to adoption of soil conservation practices.

Whereas the vast majority of rural sociological studies on soil erosion and conservation have been focused on dependent variables such as erosion rates or adoption of conservation technologies, Korsching and Nowak have been among the few to examine the correlates of farmers' views on alternative soil conservation policies. Korsching and Nowak examined farmers' attitudes toward four conservation policies: economic incentives, economic penalties, legal regulations, and educational programmes. They found that farm legal organizational complexity, total acres in crops, gross farm income, use of hired labour, and managerial complexity were positively related to favourable attitudes toward economic incentives.

None of the independent variables they examined was related to perceived acceptability of economic penalties. Use of hired labour, managerial capacity, use of credit, and use of "best management practices" were positively related to favourable attitudes toward legal regulations. Finally, use of hired labour and managerial capacity were positively related to favourability toward educational programmes. These results thus suggest that farm size generally is positively related to farmer support for the principal, widely discussed public policy strategies for achieving soil and water conservation. It should be noted, however, that these results contrast with those of Buttel et al., who found that gross farm income and number of hired workers were unrelated to concern with soil erosion among samples of Michigan and New York farm operators. They also found that gross farm income had a moderately strong inverse relationship to concern with pollution from agricultural chemicals.

The relationship between land tenure and adoption of conservation practices has been among the most inconsistent in the literature. It has long been argued that tenancy should be a barrier to soil conservation since a tenant would have little interest in the longterm productivity of land owned by someone else and because leasing arrangements are often short-term, unstable, and uncertain. Nonetheless, while some studies have reported that tenancy inhibits the adoption of conservation practices, other studies have found no such relationship. The tendency toward little or no difference in practices used on owned versus rented land probably is attributable to the fact that many farmers combine land ownership and rental and use the same practices on all of their land.

Results regarding the impact of farm debt level on conservation behaviour have also been mixed. It has been argued that high debt loads would inhibit the adoption of conservation practices, since current income is much more valuable than long-term productivity benefits for farmers facing high debt service payments and financial stress. Several studies have found support for this argument, but other studies have found no such relationship.

A much more consistent predictor of adoption of conservation practices has been the farmer's propensity to undertake risk. Nowak and Carlson et al., for example, have found a very strong positive relationship between risk propensity and the adoption and effective use of conservation tillage technology. It should be noted, however, that risk propensity is much more strongly related to adoption of conservation tillage than it is to traditional cropping practices and structures. Recent research has also found that public programmes can have significant impact on farmer soil conservation behaviour. Cost-sharing, attendance at conservation field days, having a Soil Conservation Service farm plan, and access to and use of information sources regarding conservation technologies have been found to be positively related to the adoption of conservation practices.

Two observations can be made on the empirical studies that have been reviewed above. First, these studies have generated a relatively consistent picture of the types of farm operators who are most likely to adopt soil conservation practices—namely, large, well-educated, nonrisk-averse farmers who have access to public cost-sharing and conservation information programmes. Second, these data are, in the main, consistent with the diffusion of innovations approach. But these empirical results and their consistency with the diffusion of innovations model must also be considered along with several macro-level observations that suggest limitations to this model.

One such macro-level observation is that after $15 to $30 billion of federal spending on soil conservation programmes since the 1930s, there has been no clear trend toward reduced soil erosion levels. And it has been estimated that even by 2010, only 50 to 60 percent of the nation's cropland will be farmed with minimum tillage. Thus if the diffusion of innovations approach is to be gauged, at least in part, on the basis of the efficacy of public information and subsidy programmes to reduce erosion levels significantly, critics of the diffusion approach may have a point when they argue that voluntary programmes based on influencing farmer decision-making will tend to be overwhelmed by powerful macrolevel forces leading to land degradation.

It is generally agreed that the major macrostructural trends in U.S. agriculture have been toward fewer and larger commercial farms, increased mechanization, increased specialization, and endemic overcapacity. These trends, singly and in combination, have been demonstrated to constrain soil conservation efforts. In particular, trends toward specialization as manifested in continuous cropping, cessation of crop rotations, row crop monocultures, and the separation of crop from livestock production have been found to severely exacerbate soil erosion problems. Thus, ironically, the very types of farmers who have been demonstrated to be most likely to adopt soil conservation practices are those who represent major farm structural trends that constrain soil conservation efforts. And given that farmers who have adopted conservation tillage have tended to do so for reasons other than soil conservation, one might question the permanency of the gains that have been made in reducing erosion on a voluntary basis.

Another macrostructural consideration in soil erosion research and policy concerns the cycles of overproduction, expansion, and instability that have affected agriculture in the past and are likely to do so in the future. For reasons that are largely inherent to agriculture as a production sector—high risk, the supply of land being fixed, the discontinuous nature of labour demand, and low price and income elasticities of demand for agricultural commodities—agriculture tends to have low rates of return.

The tendency toward overproduction and cost-price squeeze leads farmers to have short-term planning horizons and to heavily discount longterm returns such as those from soil conservation. And even during occasional periods, such as the 1970s, when farming becomes relatively profitable, the operation of land markets tends to discourage soil conservation practices. As farmers bid above-normal profits into local land markets and land prices increase, these higher land prices tend to encourage farmers to use land-productivity-maximizing technologies such as petrochemicals and result in exacerbation of trends toward specialization and monoculture, the principal causes of land degradation.

In sum, recent research on soil erosion and conservation has converged on a number of widely recognized generalizations, especially at the micro level of analysis. There has also been significant progress in recognizing the importance of both the micro and macro levels of analysis and in understanding their interrelations. Further work along this line can be said to represent the most promising frontier for future research on resource management and behaviour in agriculture.

The "New Environmental Debates"

Environmentally related issues have been of concern to rural sociologists not only in terms of how agricultural structures and practices affect environmental quality. Indeed, over the past few years ecological phenomena have come to be viewed as explanatory variables—and, in some formulations, as an alternative explanatory theory—in relation to agricultural structures. This work, however, has been very controversial, with much of the literature having been devoted to debates and commentaries. In this concluding portion of the chapter we will explore three of the more important debates that have ensued as a result of the development of new, innovative ecological perspectives.

The first such new environmental debate, that between Dunlap and Martin and Coughenour, was based largely on two previous ones. In the first of the previous debates, Ashby and Coward commented on an article by Gartrell and Gartrell, which was devoted to explaining adoption behaviour in the Indian state of Andhra Pradesh. Ashby and Coward noted that "there is considerable research evidence available which shows that there can be dramatic changes in farmer adoption behaviour with relatively minor differences in agrosocioeconomic conditions which affect the appropriateness of a technology to a particular farming environment." They argued that Gartrell and Gartrell failed to take these environmental factors into account despite the fact that there is considerable agroclimatic, geological, and topographic variation in the study area and suggested that the results reported might well be the spurious results of agroecological variations in the study area. Gartrell and Gartrell responded to Ashby and Coward's criticism in several respects.

First, they argued that their methodology enabled them to sidestep the issue of site-specificity for a particular innovation. Second, they argued against a principal emphasis on physical factors in isolation from the socioeconomic context of agriculture. Third, they reanalysed their data to determine if there were demonstrable effects of environmental variables and reported that "neither controls for district or village nor analysis restricted to cultivators of the same crop leads us to significantly alter our conclusions".

The second previous debate that contributed to that between Dunlap and Martin and Coughenour related to a series of studies on energy use in U.S. agriculture. Buttel and Larson used state-level census data to explore the hypothesis that farm size is positively correlated with the energy intensity of crop production. The data were found to support this hypothesis and others relating to variables such as the prevalence of legally incorporated farms. Gilles, however, was critical of the study for its failure to include environmental and climatic variables.

Gilles reanalysed the Buttel-Larson data and found that irrigation, which he considered to be an indicator of climate, was most closely related to the energy intensity of crop production and that the relationships between farm size and energy use were spurious when irrigation was controlled. Larson and Buttel replied by arguing that Gilles' irrigation variable was more an indicator of capital-intensity of agriculture than of climatic conditions. Larson and Buttel measured climate at the state level in terms of mean annual rainfall and found that inclusion of this variable in the model did not alter their conclusions. Heaton and Brown, however, explored the relationship between farm size and energy intensity by using a longitudinal analysis of county-level data and reported results largely consistent with those of Gilles.

Building on these two previous debates, Dunlap and Martin made a strong case that sociologists cannot adequately understand phenomena such as farm structure and technological change without considering the role of environmental variables. They developed a tentative model for "bringing the environment into the study of agriculture" and made several suggestions for future research directions. In particular, they argued for a conception of agricultural systems as ecological systems that involves environmental as well as social levels of causation in the analysis of farm structure and structural change.

Coughenour concedes the lack of systematic attention to such factors in most research on agriculture, but he suggests that there have been a number of recent developments that have represented major progress in incorporating social-ecological concepts. Coughenour nonetheless argues that although Dunlap and Martin correctly identify a gap in present research on the sociology of agriculture, I believe their diagnosis of the problem is only partially accurate. The main causes of neglect of the environment are twofold. Foremost is the neglect by agricultural sociologists of the process of agricultural commodity production. The main way environmental factors impinge on agriculture and vice versa are through the process of food production. It is through consideration of the inputs or resources for agricultural production that environmental variables come into view. Another reason for inattention to environmental factors is the neglect by environmental sociologists themselves of two issues: the environment as resources for the production of commodities and the effects of competition for resources on the organization of the productive process.

Coughenour then proceeds to outline a theoretical perspective on agricultural production at the farm level that includes environmental variables. He does so by:

1. Developing concepts that link fanning with the biophysical environment,
2. Cconceptualizing the interaction of farmers with the environment through the commodity production process,
3. Conceptualizing the ways that farmers deal with environmental variability and change,
4. Distinguishing between the individual and aggregate-collective levels of farmers' responses to variability and change.

A second major new environmental debate was that between Field and Johnson and Dunlap. The former argued that there has been a long tradition of rural sociological research that continues to provide a useful orienting perspective on the role of environmental factors in rural social change. They note at several points in their paper that ostensibly new "ecological" theories, such as developed by Dunlap and Martin, are not particularly novel and have, in effect, "reinvented the wheel."

Dunlap responded with a vigorous defense of his work with Martin and his related work. He stressed that while there is much of value in the older rural sociological literature, it is not consistently relevant to the study of ecological variables within a sociology of agriculture. Dunlap reiterated that sociologists have tended to ignore physical and ecological factors in their studies of agriculture and that the inclusion of such variables can greatly increase the ability of sociologists to explain certain phenomena. Dunlap contended, moreover, that the ultimate significance of an ecological approach is that it directly challenges existing paradigms, including those used by the pioneering rural sociologists, and helps to expand the boundaries of sociological discourse. Dunlap concluded by arguing that Field and Johnson have given a misleading impression of the roots of environmental sociology and have underestimated the neglect of ecological variables in recent work in the sociology of agriculture.

The third major new environmental debate has been between Albrecht and Murdock and Swanson and Busch. Albrecht and Murdock developed an ecological approach, based on a modification of human ecology, to explain the incidence of part-time farming within the universe of U.S. counties. They conceptualized part-time farming as "a form of adaptation reflecting the ability of a population of farm operators to access simultaneously two or more environmental niches reflecting farm and nonfarm forms of sustenance diversity and to compete for environmental resources". In particular, they saw parttime farming as a "relatively successful form of adaptation rather than an indication of partial failure to the changes affecting agriculture".

They found that 55 percent of the variance in the incidence of part-time farming is explained by six ecological variables percent of total acreage harvested, farm mechanization, farm diversity, nonfarm labour force participation, nonfarm technology, and nonfarm sustenance diversity), most of which is found to be attributable to the percent of total acreage in the county in harvested cropland and nonfarm sustenance diversity. They concluded that "counties with the highest prevalence of part-time farmers are those with a marginal environmental base for agricultural production and a diverse nonfarm sustenance base."

Swanson and Busch were sharply critical of the Albrecht Murdock study on several grounds. First, although they concede that ecological models may help to explain some variation in part-time farming, other conventional social and economic models are held to be much more effective predictive tools. Second, Albrecht and Murdock, in their view, have confused and failed to distinguish between the socioeconomic and natural environments, adding ambiguity to their conclusions. Third, Swanson and Busch claim that the Albrecht-Murdock model relies on cross-sectional data and cannot infer change in the "social formation" of part-time farming. Fourth, Albrecht and Murdock are said to assume a unilinear conception of social and technological change and to confuse the process by which full-time operators become parttime. Further, Swanson and Busch object to Albrecht and Murdock's portrayal of part-time farming as an adaptive process, which ignores the fact that part-time farming may represent an unstable and undesirable situation for many farm families.

Albrecht and Murdock have defended their original study against Swanson and Busch's criticism. They restate and elabourate their human ecological theory, pointing out that the original paper was intended more as a theoretical piece to stimulate discussion than a comprehensive analysis of part-time farming per se. Albrecht and Murdock respond to Swanson-Busch's criticism of their conceptual framework and methodological procedures and explore other aspects of human-ecological theory, the distinction between technological developments that involve either commensalistic or symbiotic relationships that may be applicable to aspects of farm structure other than part-time farming.

Albrecht and Murdock's response to Swanson and Busch involved a vigorous defense of the proposition that human ecological theory is useful in integrating the analysis of social and nonsocial phenomena and in researching "the adaptations of a society to a changing resource base." This posture, while having a very general similarity to that of Dunlap and Martin, is nonetheless quite distinctive in certain respects. The notion of "adaptation," which has been integral to human ecology, but also to Parsonian and other versions of functionalism, has been largely avoided in Dunlap's work. In fact, one of the key premises in Dunlap's other elaborations of his ecological perspective is that social structures and processes may tend to result in profound ecological disruption, thereby undermining the society's resource base or leading to a lack of "adaptation," to use Albrecht and Murdock's expression.

Thus, it should be stressed that the ecological perspectives that have been developed over the past five or so years are themselves quite diverse. Dunlap, in particular, has stressed that the "new environmental paradigm" involves assumptions and premises that reflect trenchant criticism of traditional sociological theory. Dunlap's criticisms presumably would extend to human ecological theory, which has been largely based on Durkheim and involves assumptions very similar to those of much structural-functionalism of the 1950s. Albrecht, Murdock, and their associates, nonetheless, must be credited with having developed a perspective that has lent itself well to systematic empirical research involving a variety of phenomena relating to the structure of agriculture.

Swanson and Busch, in a footnote to their comment on Albrecht and Murdock, noted that rural sociology has not had a tradition of scholarly dialogue and criticism. We would argue, however, that one of the most useful long-term legacies of the establishment of this new sociology (or "sociologies") of agriculture has been its stimulation of critical dialogue among researchers. Thus the emerging dialogue is multifaceted, including critical commentary among as well as within "paradigms," to use Dunlap's expression. This cannot help but be an immensely positive trend in the sociology of agriculture and in rural sociology as a whole.

Index

A

Agrarian 4, 155, 159, 165, 167, 168, 208, 257
Air bombardment 214
Albania 221, 222
Allied food 78
Alsace-lorraine 214
Animal products 210, 212, 214, 218, 225, 228, 231
Argentina 198

B

Baltic states 211, 217
Bandar shahpur 240
Basin 210, 215, 224, 234
Beef 52, 145, 147, 148, 233, 236
Belgium 209, 214, 215, 216, 222, 224, 227, 230, 231, 243
Bessarabia 210, 211, 218
Black markets 226
Blockade 211, 213, 227, 235, 243
Bohemia 227
Bolivia 192
Brazil 9, 192
Breadmaking 25
Brie-Longwy coal basin 215
Bulgaria 210, 217, 218

C

Capital 4, 8, 152, 157, 160, 162, 164, 167, 169, 171, 184, 187, 189, 246, 257, 258, 270
Caucasus 211, 238
Cereals 31, 154, 176, 203, 210, 237
Chickens 232, 233
Co-operative associations 217
Coal 210, 215
Cocoa 30
Cod-liver oil 209, 218
Coffee 25, 30, 35
Commodity cartels 216
Conferences 45
Corsica 222
Credit 27, 169, 174, 175, 176, 182, 183, 265

D

Danube 210, 218, 224, 229
Danubian countries 210, 217, 218
Danzig 215, 226
Deflation 259
Dehydration 235
Democracy 8, 206
Denmark 209, 214, 215, 216, 224, 227, 230, 231, 232, 233, 234

Depletion 234, 243
Dnieper 220, 238
Donbas 239
Dried fruits 210
Drought 3, 180, 194, 200, 229
Dumping 7, 197, 198, 199, 201, 205
Duties 137, 198

E

Employment 168
Estonia 210, 214
Europe 35, 141, 212, 219, 221, 228, 229, 231, 233, 234, 235, 238, 239, 243
European agriculture 213, 227, 235
European axis 210
Expansion of 31, 230, 257

F

Farmers 4, 5, 6, 7, 9, 10, 163, 169, 170, 171, 172, 173, 174, 178, 199, 200, 201, 202, 203, 204, 205, 206, 207, 209, 215, 217, 220, 223, 224, 225, 226, 233, 264, 266, 267, 268, 271, 273
Feed 6, 8, 18, 58, 172, 209, 212, 218, 225, 228, 230, 231, 232, 233, 236, 238, 241
Fertilizer 212, 218, 225, 233
Festung europa 227
Figs 210
Finland 210, 222, 230
Five-year plan 240
Flax 209, 211, 236
Food and Agriculture Organization 1, 7, 127
Food and nutrition 19, 39, 42, 127, 177, 188, 206
Food economy 208, 210, 211, 213, 214, 215, 216, 221, 222, 226, 231, 232
Food needs 7, 76, 181
Food research institute 229
Foreign 23, 24, 51, 135, 164, 199, 203, 221, 224, 240, 258
Foreign crops and markets 234
Friedrich 221
Friedrich naumann 221

G

Grain 6, 172, 176, 199, 201, 208, 209, 211, 218, 219, 222, 224, 229, 230, 231, 232, 234, 237, 238, 239, 240, 241, 242, 257
Grain crop 229, 232, 234, 237, 242
Great britain 221, 226, 227
Greece 210, 222

H

Holland 209, 214, 215, 216, 224, 227, 230, 231, 234

I

Invasion 210, 214, 219, 222
Iowa 259
Irrigation 6, 143, 183, 190, 270
Isolation 22, 27, 82, 269

L

Latvia 210, 214
Legislation 22, 25, 26, 27, 28, 29, 34, 41, 42, 62, 138, 183, 188, 191, 192
Lithuania 210, 214
Livestock in 234

M

Malnutrition 3, 6, 9, 33, 177
Mediterranean 222
Mittel Europa 221
Monetary 257
Monopoly 153, 190
Moravia 214, 227
Mortgages 201
Murmansk 240

N

Nazis 213, 214, 217, 219, 222, 223
Neutrals 229
North America 89, 105
Norway 156, 209, 210, 218, 227

O

Oil cake 225
Oilseeds 198, 209, 210, 211, 215, 218, 225, 229, 230

P

Paris 218
Pigs 232, 233, 241
Pork and bacon 236
Pulses 177, 202

Q

Quotas 224, 225

R

Railroads 210, 235
Reconstruction 213, 238
Revolution 21, 31, 74, 154, 181, 207
Rhine 215
Rome 7
Rubber 241
Ruhr 215
Rumania 210, 218, 230
Russia 209, 211, 214, 218, 219, 221, 223, 225, 229, 233, 235, 236, 237, 238, 239, 240, 241, 242, 243

S

Saar district 215
Safeguards against 226
Salt fish 209
Seed 10, 141, 142, 143, 178, 191, 192, 194, 195, 201, 203, 204, 206, 225, 236, 239, 241
Sheep 235, 239, 241
Sovkhoz 220
Stabilization 261
Starvation 1, 6, 200
State Department 22
Subsidies 185, 197, 198, 199, 200, 205, 224
Sudeten area 213
Sugar 21, 60, 103, 104, 105, 112, 190, 208, 209, 210, 211, 222, 230, 236, 237, 238, 242
Surpluses 210, 219
Sweden 230

T

Tallow 228
Tariff 174
Tobacco 117, 210, 218, 222
Tractors 216, 238, 242
Transnistria 219
Transportation 11, 26, 30, 36, 65, 200, 210, 215
Truck crops 230

U

Ukraine 209, 211, 214, 219, 220, 238, 242

V

Vegetables 13, 51, 61, 62, 74, 129, 130, 135, 197, 210, 222, 230, 231, 235
Venezuela 192
Vichy France 221
Vladivostok 240
Volga 211
Von Ribbentrop 221

W

Wages 167, 171, 196
War Food Administration 211, 212, 216
Weather 5, 180, 212, 228, 229, 240, 242
Whale oil 211
Whaling 211
Wheat 58, 197, 199, 203, 210, 225, 229, 230
Wood 26, 62, 218, 222
World War II 30, 31, 61, 251, 259

Y

Yugoslavia 210, 218